PHILOSOPHY DEPARTMENT LIBRARY
Waggener Hall 312
University of Texas
Austin, Texas 78712

THE JUSTIFICATION OF SCIENTIFIC CHANGE

THE JUSTIFICATION OF SCIENTIFIC CHANGE

SYNTHESE LIBRARY

MONOGRAPHS ON EPISTEMOLOGY,

LOGIC, METHODOLOGY, PHILOSOPHY OF SCIENCE,

SOCIOLOGY OF SCIENCE AND OF KNOWLEDGE,

AND ON THE MATHEMATICAL METHODS OF

SOCIAL AND BEHAVIORAL SCIENCES

Editors:

DONALD DAVIDSON, *The Rockefeller University and Princeton University*

JAAKKO HINTIKKA, *Academy of Finland and Stanford University*

GABRIËL NUCHELMANS, *University of Leyden*

WESLEY C. SALMON, *Indiana University*

CARL R. KORDIG

THE JUSTIFICATION OF SCIENTIFIC CHANGE

D. REIDEL PUBLISHING COMPANY / DORDRECHT-HOLLAND

Library of Congress Catalog Card Number 78–154739

ISBN 90 277 0181 4

All Rights Reserved
Copyright © 1971 by D. Reidel Publishing Company, Dordrecht, Holland
No part of this book may be reproduced in any form, by print, photoprint, microfilm,
or any other means, without written permission from the publisher

Printed in The Netherlands by D. Reidel, Dordrecht

TO MY PARENTS

And then there shall be in the land some who welcome him with love, who lay their hands on his head, and say: "Sit down with us to meat, live with us in our house, and share all that we have, for I have known your father".
Multatuli

When the ship was in danger, it was she who put heart into the crew, the very men to whom passengers unused to the sea turn for reassurance when they are alarmed.
Augustine

PREFACE

In this book I discuss the justification of scientific change and argue that it rests on different sorts of invariance. Against this background I consider notions of observation, meaning, and regulative standards.

My position is in opposition to some widely influential and current views. Revolutionary new ideas concerning the philosophy of science have recently been advanced by Feyerabend, Hanson, Kuhn, Toulmin, and others. There are differences among their views and each in some respect differs from the others. It is, however, not the differences, but rather the similarities that are of primary concern to me here. The claim that there are pervasive presuppositions fundamental to scientific investigations seems to be essential to the views of these men. Each would further hold that transitions from one scientific tradition to another force radical changes in what is observed, in the meanings of the terms employed, and in the metastandards involved. They would claim that total replacement, not reduction, is what does, and should, occur during scientific revolutions.

I argue that the proposed arguments for radical observational variance, for radical meaning variance, and for radical variance of regulative standards with respect to scientific transitions all fail. I further argue that these positions are in themselves implausible and methodologically undesirable.

I sketch an account of the rationale of scientific change which preserves the merits and avoids the shortcomings of the approach of radical meaning variance theorists. Radical meaning variance theorists have trouble in comparing different scientific theories. My account permits different theories to be compared in various ways. These comparisons, I argue, are possible through appeal to "first-level" and "second-level" invariance. In this connection I argue that observation, meaning, and regulative standards should be, and in fact usually are, non-trivially invariant with respect to scientific change.

This book owes much to many. It is based in good part on my Ph.D.

dissertation, *Meaning Invariance and Scientific Change* (Yale University, 1969). This dissertation has also served as a basis for the following publications: 'On Prescribing Description', *Synthese* **18** (1968); 'The Theory-ladenness of Observation', *Review of Metaphysics* **24** (1971); 'Objectivity, Scientific Change, and Self-reference', in *Boston Studies in the Philosophy of Science* (ed. by R. S. Cohen and R. Buck), vol. 8, D. Reidel, Dordrecht-Holland, 1971; 'Feyerabend and Radical Meaning Variance', *Nous* **4** (1970); 'The Comparability of Scientific Theories', *Philosophy of Science* **38** (1971). The second of these publications was the award winning entry in the 1969 Dissertation Essay Competition sponsored by the *Review of Metaphysics*. Each of these publications represent lemmas relevant to, and integrated by, the continuous chain of argument of this book.

Foremost thanks go to Richmond H. Thomason for his constructive criticism and his encouragement. His incisive and extensive comments have made this work far better than it otherwise would have been.

It was the late Norwood R. Hanson who, through his teaching and writing, first inspired my interest in these topics. What I have learned from him in and through much disagreement is beyond measure. I am also indebted to Henry Margenau and Arthur I. Fine for valuable suggestions and encouragement. What I have learned from them in and through much agreement is also beyond measure.

To Robert Stalnaker, Charles B. Daniels, Stephen Toulmin, Bas Van Fraassen, John E. Smith, Andreas Eshete, Paul Schaich, Theodore Voelkel, Arthur Sist, and Robert Kell go my thanks for valuable discussions and insights which have improved and clarified my position on many points.

I must also express my general debt to all of those mentioned above and in addition to Tony Zale, L. Michael Curtin, Robert S. Brumbaugh, Gerold Prauss, Frederick B. Fitch, Mollie Cohen, J. N. Findlay, and Robert S. Hartman both for their having sparked and sustained my interest in philosophy and for their invaluable moral sustenance. To my parents I extend my loving gratitude for their patience and encouragement.

Numbers in square brackets correspond to works in the Bibliography at the end of the book.

Acknowledgment is made to the authors, journals, and publishers for permission to quote from copyrighted materials which are listed in the bibliography.

ANALYTICAL TABLE OF CONTENTS

Preface   VII

Chapter 1: The Theory-ladenness of Observation   1

  I. Summary of the general view held by Professors Feyerabend, Hanson, Kuhn, and Toulmin: scientists who accept different theories cannot see the same things.   1

 II. I consider Hanson's position in detail. For Hanson scientists in different traditions see different things in the sense of 'see' relevant to science. I reconstruct and evaluate his argument.   3

    A. I summarize his position and reconstruct his argument. Two cases: 'sees that' means 'knows that' or 'believes that'.   3

    B. Evaluation of the first case: 'sees that' means 'knows that'. The argument is valid, but leads to absurd consequences; the premises are false. I compare the Hansonian program with the sense-data program. Other reasons suggesting the falsity of the premises: 'sees that' and 'knows that' are usually intensional; 'sees' is usually not.   6

    C. Evaluation of the second case: 'sees that' means 'believes that'. This argument is invalid. Its conclusion leads to absurd consequences: There would be a problem with rational revisions of essential beliefs. Tycho's and Kepler's beliefs would not be rival. The premises are false. Another reason suggesting their falsity: 'believes that' is intensional; 'sees' is not.   9

    D. Hanson's view of observation is incompatible with his view of retroductive inference.   11

III. I consider Feyerabend's position in detail. Scheffler's and Quine's views each illuminate the issues. One could maintain both that observational results are theory-neutral and that there are no data without concepts.   13

IV. I consider Kuhn's position in detail.   16

    A. I critically examine some of Kuhn's examples and his evidence for Gestalt shifts. 'Seeing' versus 'believing that': I sketch an alternative approach.   16

    B. An analysis of one of Kuhn's examples suggests that for Kuhn different traditions could not be rival. Source of difficulty: false or "systematically misleading" claims. I sketch an alternative approach similar to IVA.   19

  V. I carry out a general methodological *reductio ad absurdum* of the position held by Feyerabend, Hanson, Kuhn, and Toulmin.   20

    A. Revision of beliefs as to the essential properties of experience would be precluded. Progress would thus, in this sense, be precluded.   21

B. Different traditions could not be rivals or alternatives.   22
C. Kuhn, for example, seems to presuppose fixed data, namely, an environment. There is an interaction problem between theory, and environment or fact.   23
D. Observations presuppose, and are laden with, the particular theory of the time. Therefore, no theory could be tested or falsified.   25
   1. Observations and observation reports could not lead to the rational rejection of a scientific theory.   26
   2. Nor could they lead to the rational acceptance of a new theory which is inconsistent with the old.   26
VI. There is positive justification for assuming that experience is neutral with respect to alternative scientific theories. V provided us with methodological justification. The historical examples (esp. in IV) provided us with empirical support. There are two bits of further empirical evidence: (a) the existence of surprise and unsettlement; (b) scientists in different traditions sometimes use the same sorts of sentences to describe what they have observed. I examine an illegitimate use of (b) by Feyerabend to suggest the opposite of my view.   27
VII. I discuss the merits of their view and present a viable sense for 'the theory-ladenness of observation'. The confirmation and test potential of observations may change with change in theory; scientists in different traditions may therefore sometimes look for new things and sometimes in fact see new things if they find what they are looking for.   28
Conclusion: The arguments for the radical non-neutrality of observations have failed. There are, however, some merits in this view. I suggest that observation in fact is neutral. This is methodologically desirable. Science is a cumulative and expanding enterprise.   31

Chapter 2: An Examination of Some Arguments and Criteria for Radical Meaning Variance   34

I. Summary of the radical meaning variance position: this position has two central theses. The first does not, as is claimed, entail the second. I consider the relevance of some general philosophical positions concerning the theory of meaning.   35
II. The claim that modifications of a theory cause the terms occurring in it to enter different essential relations is used to support radical meaning variance. The inference from the premises to the conclusion is valid.   38
A. An examination of one of the premises: It is found to be false. A replacement is available, but it encounters another difficulty.   39
B. Two interpretations of the other premise: 39
   1. The first interpretation does not support the radical meaning-variance theorists' analyses of actual transitions.   39
   2. The second interpretation amounts to a replacement which ends up false.   40
C. The root of the difficulties of the argument discussed in II: two expressions or terms are held to have precisely the same meaning or else must be radically or completely different.   41

III. Another argument for meaning change is similar to II. However, we will weaken the conclusion; we will not interpret it as concluding that there has been a *radical* change of meaning or as assuming a distinction between essential and inessential relations. It employs a criterion to determine meaning change.   42
    A. The antecedent of the criterion requires a consistency condition. If one adds it, however, the result is merely a sufficient condition for *either* a change in the meaning of the terms *or* the mutual inconsistency of the theories employed.   43
    B. The criterion, further, leads to a logical contradiction assuming it has been used to affirm a change in the meaning of a term.   43
IV. Feyerabend has proposed a recent criterion of radical meaning change which hinges on category change and change of extension.   44
    A. The application of the criterion rests on being able to refer to unique rules which allow an unambiguous classification of the objects involved. Yet, such rules, in general, vary with the context and our purposes and are not unique. Indeed, the criterion does not support Feyerabend's conclusion that a radical change of meaning occurred in the transition from classical mechanics to general relativity.   44
    B. Further difficulties with this criterion: 46
        1. One of its implications is untenable in that a sufficient condition for meaning change entails stability of meaning.   46
        2. A more viable modification of Feyerabend's criterion. It meets my objection of IVB1 but first, does not support one of Feyerabend's historical conclusions and second, neglects operation and magnitude terms.   47
Conclusion: The proposed arguments and criteria for radical meaning variance have failed.   48

Chapter 3: The Methodological Undesirability of Adopting a Position of Radical Meaning Variance   50

The radical meaning variance position has several methodologically undesirable consequences which are not avoidable.

I. An examination of some examples, which have been put forward to illustrate and suggest the radical meaning variance position, points up difficulties. Instead of confirming, these examples instead suggest the falsity of the radical meaning variance position. Hanson's discussion of Brahe versus Kepler is incorrect for two reasons. For the same reasons similar examples adduced by Feyerabend, Kuhn, Toulmin, and Smart, are implausible.   50
II. The first methodologically undesirable consequence of the doctrine of radical meaning variance: no theory could contradict or agree with another; two different theories could be neither consistent nor inconsistent with one another.   52
    A. This consequence has revisionary, not descriptive, implications for the history of science: Bohr, Lavoisier, Priestley. Most radical meaning variance theorists claim, however, to be descriptive in such matters. Many scientists would have to be held not to have

        understood the terms they used. The consequence we draw in this section is in opposition to Feyerabend's principle of proliferation which motivates him to hold radical meaning variance in the first place. The consequence also destroys a second reason for espousing radical meaning variance.   52

  B. Neither of Feyerabend's two replies to such criticism succeeds in being able to establish a special sense of disagreement between two incommensurable theories without appealing to some shared meaning between their respective terms.   55

III. The second methodologically undesirable consequence: true communication, in any sense, between holders of different theories would be impossible. Two different theories could be neither rivals nor alternatives nor be in competition. This consequence is at odds with one of Feyerabend's reasons for espousing radical meaning variance.   58

IV. The third methodologically undesirable consequence: one could not *learn* a new theory.   59

  V. The fourth methodologically undesirable consequence: no theory could be tested or falsified by any observations or observation reports.   61

  A. All assertions of a scientific theory would, given the radical meaning variance view, be either true in virtue of the meanings of the terms employed, or presuppose the theory. In either case falsification of a theory is impossible. And in either case observation reports could not lead to the rational acceptance of a new theory which is mutually inconsistent with the old.   61

  B. These consequences are directly opposed to Feyerabend's own methodological model and to one of his principal reasons for advocating radical meaning variance.   65

  C. Kuhn presents three reasons from the behavior of scientists to the effect that these consequences would not be undesirable because testability or falsifiability is a myth. His reasons fail and his conclusion is inconsistent with the positive part of his own methodology.   66

VI. The fifth methodologically undesirable consequence: If it were true scientific change could not constitute *progress*.   70

  A. How the doctrine gives rise to this consequence. Kuhn's view provides an illustration. The consequence is also incompatible with Kuhn's positive position as to the resolution of paradigm disputes.   70

  B. Kuhn has arguments (other than those discussed in Chapter 2) which would presumably demonstrate the impossibility of scientific progress and cross-revolutionary communication. Kuhn claims that because all justifications of paradigm change involve paradigms no paradigm change can be justified. This claim is incorrect. For evaluative purposes paradigm change, contrary to Kuhn, can be viewed as a deliberative process which occurs because of features shared by competing paradigms.   73

  C. Toulmin has a (1961) redefinition of 'scientific progress' to which my claim in A is presumably not extendable. His redefinition is inconsistent with the radical meaning variance position.   74

D. Kuhn (1962), and Toulmin (1967) have another redefinition of 'scientific progress' and they advocate a purely descriptive methodology for this purpose. Their attempt fails. Among other things it is either logically untenable or else leads to an unjustified dualism. 75

VII. The sixth methodologically undesirable consequence: it is either logically untenable or else leads to two mutually incompatible dualisms each of which are unjustified and 'neo-positivistic'. 78

Conclusion: The radical meaning variance position is unreasonable. 82

Chapter 4: The Comparability of Scientific Theories 85

I. Several problems have motivated the radical meaning variance account of scientific transitions. This account does contain significant merits. Yet it also gives rise to serious problems; e.g. it would preclude the possibility of comparing different theories. 85

II. I begin an account of scientific change; it preserves the merits of the work of radical meaning variance theorists *and* permits different theories to be compared in various ways. The latter can be attained if there exists non-trivial invariance of various sorts with respect to scientific change. I examine 'first-level' invariance and argue that it in fact has occurred in scientific transitions. 89

A. Extension is a significant aspect of the meaning of a term; and if some observation is invariant then some extension is invariant. 90

B. I argue directly and indirectly that there is some non-trivial object invariance and thus some non-trivial meaning invariance with respect to several scientific transitions. Radical meaning variance theorists have denied this with respect to these transitions. In establishing object invariance I do not presuppose meaning invariance. 90

III. I sketch an account of observation in science aimed to provide a better understanding of how and why observational invariance occurs. The account I present proceeds in good part along lines suggested by Margenau. It both preserves the merits and avoids the shortcomings of the radical observational variance position of Hanson, Feyerabend, Kuhn, and Toulmin. It lends additional support to my previous claim that there usually exists some significant observational invariance and therefore some significant meaning invariance with respect to scientific transitions. On our account scientific change is a justifiable process because of invariant first-level features and elements (observation, meaning) with respect to scientific change. Our account enables us to ensure the possibility of the relation 'is a rival of' as used to compare different scientific theories. 100

IV. I suggest that rival theories can also be compared through appeal to sharable norms and a-historical standards appropriate to second-order discussion. Kuhn argues that the sharing of second-order standards is impossible. His argument is fallacious. I then briefly sketch several regulative second-order standards which are needed and used in the business of accepting, rejecting, and evaluating rival scientific theories.

XIV   THE JUSTIFICATION OF SCIENTIFIC CHANGE

I argue that each of these need not, and usually does not, change when particular scientific theories change. Taken together, first-level and second-level invariance enable us to get at the relation 'is better than' as used to compare different scientific theories.   104

Bibliography   116

CHAPTER 1

THE THEORY-LADENNESS OF OBSERVATION

Revolutionary new views concerning science have recently been advanced by Feyerabend, Hanson, Kuhn, Toulmin, and others. The claim that there are pervasive presuppositions fundamental to scientific investigations seems to be essential to the views of these men. Each would further hold that transitions from one scientific tradition to another force radical changes (a) in what is observed, (b) in the meanings of the terms employed, and (c) in the metastandards involved. In this Chapter I will discuss and evaluate (a). In subsequent Chapters I will then focus on (b) and (c) with a final return to (a).

I

Feyerabend claims that what is perceived depends upon what is believed ([16], pp. 220–221); and he maintains that among really efficient alternative theories (for the purpose of mutual criticism) "each theory will possess its own experience, and there will be no overlap between these experiences" ([16], p. 214). According to Feyerabend, "scientific theories are ways of looking at the world; and their adoption affects our general beliefs and expectations, and thereby also our experiences..." ([14], p. 29). Toulmin, Hanson, and Kuhn concur with this view. Toulmin claims that men who accept different "ideals" and "paradigms" will see different phenomena. He thinks theories not only give significance to facts, but also determine what facts are for us at all ([83], pp. 57, 95). Like Feyerabend, Toulmin asserts that "we see the world through" our fundamental concepts of science (e.g. inertial motion) "to such an extent that we forget what it would look like without them" ([83], p. 101). Indeed, both Feyerabend and Toulmin would have previous thinkers "live in an observational world very different from our own" ([16], p. 221; cf. also, [83], p. 103). Kuhn expresses quite similar views. He feels that,

... during revolutions scientists see new and different things when looking with familiar instruments in places they have looked before. It is rather as if the professional community had been suddenly transported to another planet [for] ... paradigm changes do cause scientists to see the world of their research engagement differently. In so far as their only recourse to that world is through what they see and do, we may want to say that after a revolution scientists are responding to a different world. ([42], p. 110)

This "world of their research-engagement" is similar to the objects Hanson claims scientists see in the sense of 'see' relevant to science. For Kuhn, quite literally, different "paradigms" transform observation and experience ([42], pp. 6, 110–111). "Paradigms determine large areas of experience at the same time" ([42], p. 128). He believes that after a revolution scientists work in a different world ([42], pp. 110, 117, 119, 134, 149). And one thing he means by this is that the data themselves change ([42], p. 134). This is quite different from a more traditional view. Many philosophers of science would surely want to say that what alters when a "paradigm" changes is only the scientist's interpretation of fixed data. On their view, Priestley and Lavoisier – contrary to Kuhn – both saw oxygen, but they interpreted their observations differently; Aristotle and Galileo both saw pendulums, but they differed in their interpretations of what they both had seen. For Kuhn, however, "the process by which either the individual or the community makes the transition from constrained fall to the pendulum or from dephlogisticated air to oxygen is not one that resembles interpretation" because of "the absence of fixed data for the scientist to interpret" ([42], pp. 120–121). "Normal", not "revolutionary", science utilizes interpretation because it is a deliberative enterprise "that aims to refine, extend, and articulate a paradigm that is already in existence" ([42], p. 121). During a revolution when the scientist embraces a new paradigm, he does not *interpret*. Rather, according to Kuhn, he experiences a Gestalt shift. Gestalt shifts are "sudden and unstructured", interpretation is not. Presumably there could be a time when one is half-through interpreting something, but there could not be a time when one is half-through experiencing a Gestalt shift (cf. Hanson, [27], p. 10). Kuhn maintains, along with Hanson, that the Gestalt sense of seeing is the sense relevant to revolutions in science. Normal science, on Kuhn's view, leads to the recognition of anomalies which

... are terminated not by deliberation and interpretation but by a relatively sudden and unstructured event like the gesalt (*sic*) switch. ([42], p. 121)

What were ducks in the scientist's world before the revolution are rabbits afterwards. ([42], p. 110)

Kuhn thus prefers, in analyzing scientific revolutions, to employ a conceptual scheme which has a "sudden and unstructured event" play the central role.[1]

II

A. Hanson's remarks about observation in science are indeed quite provocative. Unlike Kant he maintains that even *particular* theories determine what is seen. He admits, however, that scientists in different traditions see the same thing in one sense of 'see' ([27], pp. 5, 7, 8, 18, 20). For example, *something* about the visual experiences of Johannes Kepler and Tycho Brahe, when on a hill watching the dawn, is the same for both. Namely, both have a visual experience of a brilliant yellow-white disc centered between green and blue color patches ([27], pp. 8, 18). And for both the distance between this disc and the horizon is increasing ([27], p. 182, Note 6).

Hanson, however, claims that this type of seeing does not exhaust the concept. He feels, with Kuhn, that there is also a sense in which two observers do not see the same thing, do not begin from the same data, even though they are visually aware of the same object ([27], p. 18). He presents us with several Gestalt figures to illustrate this second sense of seeing – "seeing-as" ([27], pp. 8–18).

Well, just what have Gestalt examples to do with science? One might agree that Gestalt examples provide genuine examples of *seeing*; yet one might still doubt that observations relevant and important to science and scientific disputes are of this kind. Hanson wants to remove this doubt. He wants to claim that examples of *seeing-as* analogous to his Gestalt examples have occurred within the history of science; moreover, he wants to claim that they were more important than neutral observations for understanding scientific change and controversy.

There are then for Hanson these two senses of 'see'. In one sense (the neutral one) scientists, but presumably not *qua* scientists, see the same thing. In the other they do not. And Hanson, with Kuhn, urges it is in the latter sense, that we get *scientific* 'data' ([27], pp. 4–5, 17; [42], p. 134). Thus, he concludes that Tycho and Kepler do not begin their inquiries from the same data, do not make the same observations, do not even see the same thing ([27], p. 5). Together on a hill Tycho and Kepler, according to Hanson, see (in the non-neutral sense of 'see') different things in the east at dawn. How could it be otherwise? "Practicing in different worlds, the two ... scientists see different things when they look from the same point in the same direction" (Kuhn, [42], p. 149). "Kepler and Tycho are to the sun as we are to Figure 4, when I see the bird and you see only the antelope" ([27], p. 18).

A difficulty arises here. Hanson says you and I *both* see his figure 4, although we don't both see a bird. If then the sun is to be equated with figure 4, as Hanson has just suggested, *both* Tycho and Kepler should see it. Hanson's analogy, if accurate, seems to undercut his own position; he

Fig. 4

maintains that it is the sense in which Tycho and Kepler do *not* observe the same thing which must be grasped if one is to understand fundamental disagreements within science.

Let us, however, proceed to examine Hanson's argument that Tycho and Kepler see different things in a sense important and relevant to science.[2] The argument starts from two virtually incontestable premises,

(1) m_0 (Tycho) sees X_0 (the sun Tycho sees).
(2) m_1 (Kepler) sees X_1 (the sun Kepler sees).

Now, for Hanson, to *see* an object X is to *see that* if $A_1, \ldots A_n$, were done to X then $B_1, \ldots B_n$ respectively would result ([27], pp. 20–21, 22–23, 29–30, 58–59, 97).[3] In short, it is to see that X behaves in certain essential ways, i.e. has certain deep dispositional properties. For one to see Tycho's sun (X_0) is for him to at least see that from some celestial vantage point the sun is such that it could be watched circling our fixed earth ([27], p. 23). That is, he would see a sun which is essentially mobile ([27], pp. 17, 23–24, 182). "Watching the sun at dawn through Tychonic spectacles would be to see it in something like this way" ([27], p. 23). Hanson, therefore, adds (3) as a further premise to his argument.

(3)   If anyone sees X_0 then he sees that $P_0(X_0)$.

where 'P_0' stands for 'is mobile', a predicate in Tycho's system which holds essentially of the sun.

The situation is, however, different for Kepler's sun. For one to see Kepler's sun (X_1) is for him to at least see that the sun does not behave in the above "Tychonic" (cf. [27], p. 23) way. Why? Well, Kepler sees a static sun, whereas Tycho sees a mobile sun ([27, pp. 17, 23–24, 182). When Kepler sees the sun he sees the horizon dipping or turning away from it. When Tycho sees the sun he sees it ascending or rising ([27, pp. 23, 182). "The shift from sunrise to horizon-turn is analogous to the shift-of-aspect phenomena [viz., the Gestalt examples] already considered..." ([27], pp. 23–24). That is, Hanson adds another premise, (4), to his argument.

(4)   If anyone sees X_1 then he sees that $P_1(X_1)$.

where 'P_1' stands for 'is static', a predicate in Kepler's system which holds essentially of the sun.

From (1) and (3), Hanson validly infers (5),

(5) m_0 sees that $P_0(X_0)$.

And from (2) and (4), he similarly infers (6),

(6) m_1 sees that $P_1(X_1)$.

Now, m_0 and m_1 presumably know that anything that is static is not mobile. Therefore, Hanson obtains (7) from (6),

(7) m_1 sees that $\sim P_0(X_1)$.

From (5) and (7) he then concludes (8),

(8) $X_0 \neq X_1$.

(8) says Tycho and Kepler do not see the same sun. But the validity of the inference from (5) and (7) to (8) is questionable. It depends on the meaning of 'sees that'. We will examine this shortly. Let us for the moment assume it to be valid. Then one link remains in Hanson's argument. Why is this sense of 'see', and not the neutral sense, important and relevant to Kepler's and Tycho's observations? Simply because the differences in what they see amounts to the difference between an essentially geocentric and essen-

tially heliocentric universe. And this *is* the essential and deep difference between Tycho's and Kepler's physical theories.

The preceding is Hanson's argument. I feel it is fallacious. Hanson implicitly uses the phrase 'sees that' to mean either 'knows that' (cf., e.g., [27], pp. 18, 20–22) or 'believes that' (cf., e.g., [27], pp. 23–24). In either case his argument is unacceptable.

B. Consider the case in which 'sees that' means 'knows that'. In this case Hanson's argument is valid, but leads to absurd consequences; and we shall discover that this is because its premises are false. From (5) and the fact that we are taking 'sees that' to mean 'knows that' we can infer (9),

(9) m_0 knows that $P_0(X_0)$.

Now, whatever else it may be, what is known is at least true. That is, if m_0 knows that $P_0(X_0)$ then we can conclude that $P_0(X_0)$ is the case. Therefore (10) follows,

(10) $P_0(X_0)$.

From (7) we can similarly infer (11),

(11) $\sim P_0(X_1)$.

To show Hanson's argument is valid we will show that (8), his conclusion, follows from (10) and (11). And this is readily done: for if $X_0 = X_1$, by (10), we would have $P_0(X_1)$ which contradicts (11). Thus, Hanson's argument is valid.

It is, nevertheless, fallacious. The careful reader may have already noted the oddity of (10). Yet (10) was needed to make Hanson's argument valid; and it was indeed obtainable on the supposition that 'sees that' means 'knows that'. (10) *is* strange. Tycho's theory is false. He was mistaken about the status of the sun in the solar system. He mistakenly thought it was mobile. Either we hold (10) or not. If we hold (10) we will have to give up the widely accepted view that Tycho was wrong in thinking his sun revolved around the earth. On the other hand, if we give up (10) we will be forced to give up either (1) or (3). They were the premises which entailed (10). One could not, therefore, rationally hold them and deny (10). Now, the truth of (1) is, I think, beyond question. To deny it amounts to the claim that Tycho didn't see the sun he saw, which is absurd. If 'sees that' means 'knows that' Hanson's premise (3) should, therefore, be abandoned if we deny (10).

Hanson would, given his philosophical position, have to maintain that a geocentric world picture really isn't wrong after all. However, Hanson's own point of view is also that of modern science. Thus, he should say that a geocentric world picture is wrong, and indeed was wrong before Kepler. He should therefore maintain that a geocentric point of view is both wrong and not wrong if his own point of view includes both modern science and his own philosophical position. At the expense of the viewpoint of modern science he might maintain that his philosophical position is the single, preferred, or absolute point of view. But then he would be open to the charge that, in spite of his explicit disclaimers, he is not engaging in a *descriptive* philosophy of science; he would be, rather, philosophically legislating to science – telling scientists that what they regard as correct scientific beliefs (e.g., that the solar system has never been geocentric) are, in fact, wrong. And this charge *Hanson* would have to take seriously. He feels that a principal merit of his own approach is its descriptivism and its avoidance of the normative excesses of positivism.

Further, there are reasons which tend to suggest the falsity of many claims that, *prima facie* at least, are similar and analogous to (3) and (4). The Hansonian enterprise is not unique. Many philosophers have tried to thread knowledge into what they came to call 'sense data'. In the philosophical literature on perception and knowledge sense data of various sorts have been proposed in order to function as the basic perceptual objects. The tendency in sense data accounts is to attempt to find or to postulate some type of object the perception of which is *sufficient* for knowledge of the *seeing-that* or *knowing-that* kind. Hanson's search for the basic perceptual objects of science the perception of which is sufficient for knowledge of the *seeing-that* or *knowing-that* kind is thus, in spite of his claims to the contrary ([27], p. 22), not so different from the approach of the sense data theorists. Similarly, Hanson thinks that " 'Seeing that' threads knowledge into our seeing" ([27], p. 22), and that it is "the logical element which connects observing with our knowledge" ([27], p. 20). Daniels ([10]; [11]) and Sellars ([74]) have argued with respect to the sense data positions that none of the things customarily labelled 'sense data' guarantee that the person who sees them will know that or see that they have attributes or qualities of a particular sort. Daniels, for instance, claims that one can still veridically see any of the customary types of sense data and be mistaken about what color it is, what shape it is, etc. ([10], pp. 4–5). Daniels' argument hinges on the fact that *seeing-that* and *knowing-that* are for the most part intensional whereas *seeing things or events* is for the most part not. I can see that a thing is a lamp without seeing that it is a collection of atoms and without seeing that it is my maiden aunt's favorite possession. *Seeing-that* usually has more to do with *knowing-that*, I think, than does seeing things. One cannot see that a thing is a lamp without knowing that it is. But one can see one's maiden aunt's favorite possession without knowing that it is one's maiden aunt's favorite possession. Veridical seeings of things often should not, therefore, be taken as sufficient conditions of *seeing-that* or *knowing-that*.

This might be thought puzzling in view of the fact that we often answer such questions as "How do you know that the tallest girl in Asylum High School was at the play?" by saying "I saw her there". But suppose that the tallest girl in Asylum High School also happens to be the youngest Smith girl and that one does see the tallest girl in Asylum High School there. If this is so, one also sees the youngest Smith girl there. How, then, by seeing the youngest Smith girl there, does one end up knowing that the tallest girl in Asylum High School was at the play, but not knowing that the youngest Smith girl was? This would, I think, be indeed paradoxical if one takes veridical seeings of things cited in such answers as "I saw her there", as sufficient conditions of *seeing-that* or *knowing-that*. The proper conclusion to draw, therefore, seems to be that veridical seeings of things, or events, or sense data of the sorts discussed by Daniels, often are not sufficient conditions of *seeing-that* or of *knowing-that*.

Although this conclusion does not *establish* the falsity of Hanson's premises, (3) and (4), it does tend to make them *somewhat* implausible for each of his premises also claims that *seeing* is a sufficient condition for *seeing-that* or *knowing-that*. One might here object by claiming that (3) and (4) are special and that their truth instead hinges on the fact that P_0 and P_1 are *essential* properties of X_0 and X_1, respectively. This, however, will not do and not merely because of Quinian reasons. At any given time an object may have many essential properties which are unknown. Yet this cannot reasonably be said to prevent us from seeing the object. That we know an object's essential properties is not, therefore, a necessary condition of our seeing the object.

It is interesting to note that what we have said here is not opposed to a Kantian position. Kant would *not* have held that we cannot experience things unless we *know that* or *see that* they presuppose *his* categories; otherwise there would have been no point to his writing the *Critique*.

But if the role of such answers as "I saw her there" is usually not one of logical sufficiency for *seeing-that* or *knowing-that*, then what is its usual relation to these latter concepts? Austin perhaps provides an answer: by saying that one sees a thing or event one indicates to one's hearer that one is in a position here and now *to see that* it has attributes of a particular sort (to see that the tallest girl in Asylum High School is in the audience) ([5], pp. 47–50). But the fact that one can see the youngest Smith girl in the audience and not see that the youngest Smith girl is in the audience shows that much more than seeing x is relevant to one's seeing that $F(x)$.

C. Let us now consider the case where 'sees that' means 'believes that'. In this case Hanson's argument again is fallacious. (5) and (7) become (12) and (13) respectively,

(12) m_0 believes that $P_0(X_0)$.
(13) m_1 believes that $\sim P_0(X_1)$.

But to conclude (8) from (12) and (13) is invalid. m_0 believes that P_0 holds essentially of X_0. m_1 believes that $\sim P_0$ holds essentially of X_1. But this does not imply (8). A counterexample arises from noting the consistency of (14), (15), and (16),

(14)   Mohammed X, a Black Muslim, believes that Mohammed Ali is good, and essentially so.
(15)   Adolph Übermensch, a Nazi, believes that Cassius Clay is not good, and essentially so.
(16)   Yet, Mohammed Ali = Cassius Clay.

Hence, two people can believe that contradictory properties hold essentially of the *same* object. Thus, when 'sees that' means 'believes that' Hanson's argument is invalid.

In addition (3) and (4) become implausible. They would, if true, yield consequences which are both strange and inconsistent with other parts of Hanson's approach. If (3) were true, Tycho could not rationally revise his essential beliefs about the sun he sees and still be able to see it. Let us assume that Tycho sees the sun X_0 at time t_0. Therefore, by (3) with 'believes that' substituted in place of 'sees that', Tycho at t_0 believes that $P_0(X_0)$. That is, Tycho at t_0 believes that the sun is essentially mobile. The moment t_1 (here t_1 is later than t_0) Tycho rationally believed that $\sim P_0(X_0)$ it would follow that he would no longer be able to see X_0 if (3) were true: assuming he did see X_0 at t_1 then, by (3), he would believe that $P_0(X_0)$; and this is contrary to our previous assumption that he rationally believed that $\sim P_0(X_0)$. To believe both that $P_0(X_0)$ and that $\sim P_0(X_0)$ is not rational. Tycho could not therefore revise his beliefs about X_0. (4) results in an analogous consequence. In short, it would be impossible for (say) Tycho, in a rational way, to believe or learn that the sun he saw before is not mobile if he was still able to see it. This is paradoxical, for, in an important sense, scientific beliefs are about matters-of-fact (data, phenomena, etc.). Hanson would not disagree with this; he too would maintain that scientific beliefs express the connections thought to obtain between matters-of-fact. What then about scientific disputes and disagreements? We can grant that scientific disputes are usually not disputes over facts in the sense that courtroom disputes often are (cf.: Wilson: 'Yes, I saw Jones kill Smith'; Benson: 'No, Jones was having dinner with me when Smith was killed'). Yet scientific disputes are disputes over facts in another sense. They involve scientists' adherence to incompatible beliefs *about* facts. Such beliefs express the different connections that each thinks obtains between matters-of-fact. This point too Hanson has noted well:

The difference is not about what the facts are, but it may very well be about how the facts hang together. ([27], p. 118)

Given that Hanson (correctly, in my opinion) maintains this – and it is a central tendency in his approach – it is hard to see how the rest of his view would make it possible for scientific disputants to revise their beliefs as to the connections they think obtain between objects which they could continue to experience. In short, it would be impossible for Tycho, in a rational way, to believe, or learn, that a sun which he could continue to see throughout the time interval $t_1 - t_0$ was not mobile. If Tycho, in a rational way, believes that $\sim P_0(X_0)$ at t_1, then Tycho could not see X_0 at t_1. Granted, if he were to revise his beliefs in this way, he would see another sun X_1 (where $X_1 \neq X_0$). At t_1, however, his new and revised belief $\sim P_0(X_0)$ would not be about X_1. '$\sim P_0(X_0)$' would be about X_0. Generalizing, this would imply that one's *revised* scientific beliefs are at any time not *about* what one can see *at that time*.

This is paradoxical. It is also, as we noted, opposed to the tendency in Hanson's position which regards scientific beliefs as expressions of the connections thought to obtain between the elements of possible experience and which regards scientific disagreements as differences "about how the facts hang together."

There is another reason which would tend to suggest the implausibility of (3) and (4) when 'sees that' means 'believes that'. *Believing-that* locutions are for the most part intensional; *seeing things or events* locutions are for the most part not. I can believe that a thing is a lamp without believing that it is a collection of atoms and without believing that it is my maiden aunt's favorite possession. But, if I see a lamp, I see a collection of atoms; and if the lamp is my maiden aunt's favorite possession, I see my maiden aunt's favorite possession. Thus, one can see a collection of atoms without believing that it is a collection of atoms; one can see one's maiden aunt's favorite possession without believing that it is her favorite possession. Veridical seeings of things should not, therefore, usually be taken as sufficient conditions of *believing-that's*.

The situation is even worse. The very conclusion of Hanson's argument leads to untoward consequences. (8) tells us Tycho and Kepler each observe different things in a sense. And Hanson's conclusion is that this sense is the one relevant to their scientific disagreements. But given this it becomes impossible for them to have scientific disagreements about the sun; they are not talking about the same sun because in the sense relevant to science they don't see the same sun. And each is talking about what he sees. Their beliefs about the sun are not, therefore, *rival* beliefs. If I observe a bird and claim it can fly while you observe an antelope and claim it cannot, we are not expressing *rival* beliefs. We are seeing different things; thus, our respective claims about them are not *rival* claims. Just so, if Tycho sees something different than Kepler, then the beliefs they express about what they see are not *rival* beliefs. In short, the sense in which Tycho and Kepler do not see the same thing is *not* relevant to their scientific disagreements if these disputes are indeed disputes "about how the facts hang together."

D. Hanson suggests that observations are theory-laden, that different

scientists see different things because they espouse different theories, that
they could not have observed what they did if they did not hold the theory
that they did. This, however, is incompatible with a principal tendency of
the rest of his approach, namely, the stress he places on the importance
of "retroductive inferences" to science. Indeed, his view as to observation
would preclude the possibility of such inferences.

Hanson maintains that retroductive inference is central to science. He
thinks its form is the following:

1.   Some surprising phenomenon P is observed.
2. P would be explicable as a matter of course if H were true.
3.   Hence there is reason to think that H is true. ([27], p. 86)

Hanson thinks that:

From the observed properties of phenomena the physicist reasons his way towards
a keystone idea from which the properties are explicable as a matter of course. ([27],
p. 90)

Hanson's position on observation, if correct, however, would make it
impossible for the scientist to reason in this way. We could not begin with
data and search for the keystone principles which explained them if, as
Hanson maintains, we could not observe this data unless we already *saw
that* these keystone principles were correct (cf. "The knowledge is there in
the seeing" [27], p. 22). One could not begin with an observed phenomenon
P and search for the as yet unknown 'keystone idea' or 'hypothesis' which
would explain it. For if we could, this would imply that we do not *see that*
the keystone idea or hypothesis H is correct at the time we initially observe
the phenomenon P. Thus we could not have observed P to begin with for
the requisite knowledge (H or the keystone idea) would *ex hypothesi not*
have been "there in the seeing" at the outset ([27], p. 22; cf. also [27],
pp. 20–21, 26, along with (3) and (4) above). Retroductive inferences would
therefore be impossible. Consider Kepler. Initially he did observe a sun
P_0. But *initially* he did *not* see that this sun was static (H). Only later did
he reason his way towards this conclusion and it was a conclusion in part
based on his *(and Brahe's)* observations of the sun. But if (4) is correct
then the second he saw that the sun is static his observation of it would
have changed ("the shift-of-aspect phenomena … is occasioned by differences between what Tycho and Kepler think they know" ([27], p. 24)).
And it is this new sun P_1 which he sees that is supposed to be explicable by,

and structured by, *H*. *H* patterns, structures, and explicates P_1, not P_0. It is the sun Kepler sees at the time he *sees that H* which is supposed to be explicable by means of *H*. Thus, Kepler really would not have found an explanation for his initial observation P_0. His inference therefore would not have been "retroductive" given Hanson's account of such inferences.

III

Let us next turn to Professor Feyerabend. Within the "logical empiricist" tradition accounts of experience usually maintain that neutral observations are possible and that observational results can "be stated and verified independently of the theories investigated" (cf. Hempel, [31], pp. 20–22, 43–44). Feyerabend, however, feels that:

> In these accounts it is taken for granted that observational results can be stated and verified independently, at least independently of the theories investigated. This is nothing but an expression, in the formal mode of speech, of the common belief that experience contains a factual core, that is independent of theories.... ([16], p. 151)

Feyerabend seems to maintain, with Kant, that there are no data without concepts. For him, the given is a myth. Further, he would seem to argue, with Kant, that *even if there were such data,* "Intuition without concepts is blind." That is, they would exhibit a radical poverty of content which would preclude their epistemic relevance to the testing of scientific theories ([16], pp. 194–197, 202–215). These claims, however, do not imply, as he seems to think, that neutral observations are not possible or that observational results cannot "be stated and verified independently of the theories investigated." Affirming the possibility of neutral observations and the possibility of observational results being stated and verified independently of the theories investigated is logically weaker than affirming an unconceptualized given; it is not the case that the former are "nothing but an expression in the formal mode of speech of the common belief that experience contains a factual *core* (my italics)..." Scheffler has recently seemed to argue just this ([73], Chapter 2). He argues persuasively against Lewis' notion of the given ([46], esp. Chapter 2) and would agree with Feyerabend that pure data are a myth. However, Scheffler maintains that testing of theories does *not* presuppose unconceptualized data. He feels that observation is conceptualized. But this, he claims, is only to say that concepts (or categories) are employed to classify and individuate phenomena. Hypotheses, on the other hand, are formulated concerning the distribution of phenomena in the categories employed. The correctness of those hypotheses is tested by observation.

Feyerabend *et al.* might object that when scientists have formulated incompatible theories concerning a particular realm of phenomena, observation cannot provide an independent basis for deciding between the theories. If observations are structured by concepts, then scientists advocating distinct theories will structure their observations differently. Thus, each scientist will make observations relevant only to his own theory.

Scheffler feels his distinction between categories and hypotheses would provide an answer to such an objection ([73], p. 43). Using one basic set of categories, any number of alternative hypotheses may be formulated. Competing scientific theories are often alternative hypotheses formulated, on Scheffler's view, using one basic set of categories. This point is moot (cf. Chapter 2, Section IV). In addition, however, Scheffler feels, correctly in my opinion, that *even if* two scientists employ "non-overlapping category systems" (presumably, systems in which no category of one is identical to any category of the other) their respective categories *may* "have in common certain identical items, although they group them differently" ([73], p. 41). His example of two non-identical categories, "giraffe" and "animal" presumably supports this. Thus, he maintains that "Two category systems, as wholes, may even have all their items in common, though classifying them in different ways" ([73], p. 41).[4] For these reasons Scheffler holds that scientists advocating alternative theories often can invoke the same observations as relevant to testing both (or all) the competitors and as providing a basis for deciding which is more adequate.

Feyerabend *et al.* could perhaps respond to this argument in a manner not unsimilar to Quine's thoroughgoing pragmatic approach. Feyerabend might well deny the validity of the distinction between (a) concepts, definitions, and analytic truths, which determine the basic scheme of categories to be employed, and (b) the hypotheses or general empirical truths. He might grant that in a scientific theory or language we can make an abstract distinction, for example, between (a) general conceptual truths and (b) general empirical statements (hypotheses); yet he would perhaps deny that we can, in studying the basis for rejecting or modifying scientific theories, distinguish between those considerations which lead us simply to reject a hypothesis and those considerations which lead us to reject a system of categories. He might hold that the total theory consisting of the system of empirical hypotheses *and* the "analytic" statements which formulate concepts faces the test of observation as a loosely integrated unit. In adjusting the theory to accommodate observations, he might continue, there is a great deal of flexibility in regard to which hypotheses and statements are allowed to be altered. There is, Feyerabend might contend, no valid distinction within statements of a theory between those which formulate concepts or categories and those which express empirical hypotheses.

Such a reply is, it seems to me, unsatisfactory. It would at this point perhaps be illuminating to briefly note a *prima facie* similarity and a *prima facie* difference between Quine's and Feyerabend's approach. Both would seem to deny there is an analytic-synthetic distinction, and not merely in the sense that there is no sharp distinction. As Harman points out ([29], pp. 125–127), for Quine this denial takes the form of a claim that nothing is analytically true, that all truth or falsity is synthetic; for Quine, the analytic-synthetic distinction is "meaningless" in that one cannot make sense of it in any way such that there turn out to be analytic truths. But for Feyerabend, as we shall argue in Chapter 3, this denial should take the form of a claim that nothing is synthetically true, that all truth or falsity is linguistic; for Feyerabend, the analytic-synthetic distinction is "meaningless" in that one cannot make sense of it in any way such that there turn out to be synthetic truths; the "truths of language – truths of fact" distinction resembles the nonwitch-witch distinction, which fails to distinguish anything since there are no witches.

The basic conflict, however, between the defender of the possibility of common observations, and his critic need not, at this point, hinge either on the distinction between concepts (categories) and empirical hypotheses or on the Duhem-Quine Thesis. We might agree that both categorical schemata and particular empirical hypotheses in the form of expectations and beliefs influence both observational mode and content. Nevertheless, this does not compel us to assert that observational mode and content cannot be *shared* by holders of different empirical hypotheses. The fact that what is observed is influenced by belief does not imply that what is observed cannot be shared by holders of different beliefs; not every influence need be a one-to-one influence. Further, there may exist much in common between the two sets of beliefs and this may make common observations possible. The gap in scientific transitions is much exaggerated, as it always is in revolutionary changes; if, in any given "scientific revolution", a list were made of the agreement of the two sides, the list of agreements would be enormously large. It is a historical truism that revolutions change much less than they seem to on the surface, and this may be applied to the problems we are discussing.[5] In a conflict between competing scientific theories there are normally many rather basic principles held in common by advocates of each of the competing views.[6] Such common principles might even be said to form a system of concepts within which the conflicting hypotheses can be formulated.

IV

Kuhn's overall view of observation is quite similar to Feyerabend's and Hanson's, as we have already indicated (I). Let us, however, examine some of the specifics of his position.

A. Kuhn feels that Sir William Herschel's discovery of the planet Uranus provides an example of a celestial body that was seen differently ([42], pp. 114–115). He thinks that this discovery was a prime example of a transformation in the scientists' field of vision, one quite analogous to a Gestalt switch. His evidence for this conclusion is the following: On at least seventeen different occasions between 1690 and 1781 several of Europe's most eminent observers "had seen a star in positions that we now suppose must have been occupied at the time by Uranus" ([42], p. 114). "One of the best observers in this group had actually seen the star on four successive nights..." ([42], p. 114). Herschel "observed the same object" but after further investigation "announced that he had seen a new comet!" ([42], p. 114). Only later did Lexell suggest that the orbit was probably planetary. From this evidence Kuhn concludes,

When that suggestion [Lexell's] was accepted, there were several fewer stars and one more planet in the world of the professional astronomer. A celestial body that had been observed off and on for almost a century was seen differently.... ([42], pp. 114–115)

This same type of argument is used again and again by Professor Kuhn. For example,

...after the assimilation of Franklin's paradigm, the electrician *looking* at a Leyden jar *saw* (my italics) something different from what he had seen before. The device *had become* (my italics) a condenser.... ([42], p. 117)

Again,

Lavoisier... *saw* oxygen where Priestley had *seen* (my italics) dephlogisticated air and where others had seen nothing at all. ([42], p. 117)

Pendulums too are described as involving a "shift of vision" ([42], p. 118). Kuhn claims,

Pendulums were *brought into existence* (my italics) by something very like a paradigm-induced gestalt switch. ([42], p. 119)

Let us now evaluate these remarks. The historical evidence to the effect that the discovery of the planet Uranus was an instance of a Gestalt-type shift is dubious. It would make much better sense to say that the astronomer-observers mistakenly believed that what they had seen in the sky was a star, whereas it really was a planet. The fact is that Herschel didn't claim Uranus was a comet until *after* some further scrutiny ([42], p. 114); after this scrutiny he "announced that he had seen a new comet" ([42], p. 114). But *announcing after* investigation, that what one had previously seen is a comet is surely different from *actually seeing* something different. What Herschel did was simply announce that he had some justification for believing that what he saw was a comet. That is, Herschel announced that he had some justification for believing that what he saw possessed the dispositional properties that comets possess. Lexell suggested that the orbit was "probably planetary" "after fruitless attempts to fit the observed motion to a cometary orbit" ([42], p. 114). Surely what this suggests is *not* that Lexell now *saw* a planet ([42], p. 114) or now *perceived* differently ([42], p. 116) or now experienced a transformation of *vision* ([42], pp. 110, 119); rather it suggests that *because of* theoretical considerations Lexell now *believed that* the "observed motion" was that of a planet. That is, Lexell now rationally believed that the same object everyone had *seen* possessed the property of being a planet. He believed that it behaved in planetary ways.[7] Not *everyone*, of course, believed that what they saw was a planet. Some were mistaken about the dispositional properties what they saw possessed. Some believed that it was a star. And believing that it is a star is quite different from actually seeing or perceiving a star (cf. Section II C above). If "there were several fewer stars and one more planet in the world of the professional astronomer" ([42], p. 114) *after* Lexell's suggestion was accepted, as Kuhn claims, then something else must also follow. It must follow that *before* Lexell there were more stars in the world of the professional astronomer. But if this is true one can no longer say that these astronomers before Lexell were *mistaken* about the number of stars. Even if one held that to say whether they were mistaken is to say something about their viewpoint from one's own viewpoint, Kuhn would have to hold they were not mistaken. This is because from *Kuhn*'s viewpoint there really were more stars in their world. It is not just that they believed there were more stars in their world. According to Kuhn there *really* was this number of stars ([42], p. 114). Thus these astronomers were not wrong after all. However, Kuhn's own point of view is also that of modern science. Thus, he should also say that the pre-Lexell framework is wrong, and indeed was wrong before Lexell. He should therefore maintain that the pre-Lexell framework is both wrong and not wrong if his own point of view includes both modern science and his own philosophical position. Kuhn would thus end up espousing a contradiction unless he were to give up either his philosophical point of view or his scientific point of view. At the expense of the viewpoint of modern science he might maintain that his philosophical position is the single, preferred, or absolute point of view. But then he would be open to the charge that, in spite of his explicit disclaimers ([42], pp. 93, 161, 165, 169), he is not engaging in a *descriptive* philosophy of science; he would be, rather, philosophically legislating to science – telling scientists that what they regard as correct scientific beliefs are, in fact, wrong. And this charge *Kuhn* would have to take seriously. He feels that a principal merit of his own approach is its descriptivism and its avoidance of the normative excesses of "logical empiricism".

The same holds for Kuhn's other examples. After the assimilation of Franklin's paradigm the electrician looking at a Leyden jar did *not* see something different. The device did not *become* a condenser. Rather, it was now *believed* (or known) to possess new properties, condenser-properties. Different features of what was seen were now held or believed to be more important. Their importance was due to their now more central relation to the new conceptual framework. Shifts of vision *are* shifts of *vision*, not belief. No profit results from an inordinate intertwining of these concepts. Sight is one thing, belief another. The same holds for dephlogisticated air versus oxygen, and falling stones versus pendulums.

These paradigm shifts involved, and were due to, shifts of belief, not vision. (Hanson argues that different interpretations are not the cause of rival scientific accounts. He argues rather that disparities arise because the fundamental visual data have changed, ([27], pp. 8–11). Some of these arguments are irrelevant. They are directed at *Gestalt* examples. And this is the very question at issue. Are Gestalt shifts central to what goes on during *scientific* change?)

B. Further, Kuhn's view, as does Hanson's (cf. IIC), implies rival traditions are not really *rival*. This is because they are about different subject matters. We will soon argue this in general for Kuhn, Toulmin, and Feyerabend. Now, however, let us examine one of Kuhn's examples. In describing a famous debate between the French chemists Proust and Berthollet, Kuhn says,

The first claimed that all chemical reactions occurred in fixed proportion, the latter that they did not. Each collected impressive experimental evidence for his view. Nevertheless the two men *necessarily* (my italics) talked through each other, and their debate was entirely inconclusive. Where Berthollet saw a compound that could vary in proportion, Proust saw only a physical mixture. To that issue neither experiment nor a change of definitional convention *could* (my italics) be relevant. The two men were as fundamentally at cross-purposes as Galileo and Aristotle had been. ([42], p. 131)

Notice how Kuhn's position entails that these two men *necessarily* talked through each other, that they were *necessarily* at cross purposes. Perhaps they were, in fact, at cross purposes. Perhaps they did, in fact, talk through each other. But why necessarily? Kuhn's methodology and radically idealist theory of perception alone is what forces us to this untoward conclusion. *His* view indeed implies that they were necessarily at cross purposes, that they necessarily talked through each other. For to the extent they were brought up and trained in a given, particular tradition, they had no control over what they saw. Given that one is practicing *within* a given tradition one either sees things or one doesn't. Given then that the visual worlds of these scientists was different, as Kuhn maintains, then the claims they make about their respective worlds are not *rival* claims. The claims, instead, will be at cross purposes. When they make the claims they will indeed be talking through one another. And necessarily so, given they espouse different paradigms. But just what leads to these undesirable conclusions? I feel it is in part Kuhn's strange use of words, his strange and misleading locution, that is responsible. Language can mislead. And it in good part is responsible for the odd results we have just noted. Precisely what is responsible is the following claim which, in my opinion, is either false or misleading:[8]

> (17)   Where Berthollet saw a compound that could vary in propor-
>   tion, Proust saw only a physical mixture. ([42], p. 131)

Why not simply assert (18)?

> (18)   Berthollet believed that chemical reactions involved com-
>   pounds that could vary in proportion, whereas Proust be-
>   lieved that chemical reactions involved only physical mixtures.

Any "talking through" one another at "cross purposes" that might have gone on would, if we hold (18) instead of (17), be due to the men themselves, and not to the nature of *science*, *qua* science, as Kuhn would have it. Men do have control over what they *believe* about what they see; at least more control than over what they see. And if *these* men didn't then it was due to their own dogmatism, not to any defect of, or constraint on, their vision.

V

More general, yet analogous, objections hold against Feyerabend, Hanson, Toulmin, and Kuhn. Each of their views (cf. I), I shall argue, prevents scientists within a tradition from revising their beliefs as to the "essential" properties of experience. Progress could not, therefore, be achieved. Second, I will maintain that each of their views prevents the scientific theory after a revolution from being a rival or alternative to the scientific theory before the revolution. Third, I will argue that their views are beset by problems as to how a theory interacts with "the common environment" to produce the world. Fourth, I will argue that if their views were correct then no theory could be tested or falsified by any observations; observations and observation reports could not lead to the rational rejection of the particular scientific theory which is employed; nor could they lead to the rational acceptance of a new and revolutionary scientific theory. Each of these four conclusions constitutes some grounds for doubting the positions which lead to them.

A. Let us proceed with the first point. Roughly put, essential revisions of beliefs about experience could not be made in a rational way if Feyerabend, Hanson, Toulmin, or Kuhn were correct. This is analogous to the problem that we have already noted with regard to the details of Hanson's position (cf. IIC). There we argued that (3) and (4) were implausible (cf. pp. 9–11). If they were true, Tycho and Kepler could not rationally revise their essential beliefs about the sun they see and continue to be able to see it. Generalizing, Hanson's position as to observation would imply that a scientist's revised, essential (or "important" if one prefers to eschew essentialism) beliefs are at any time not about what he could see *at that time*. The concept of scientific progress would therefore be attenuated.

The same type of argument can readily be extended to Feyerabend, Toulmin, and Kuhn. Feyerabend says that among really efficient [9] alternative theories "each theory will possess its own experience, and there will be no overlap between these experiences" ([16], p. 214). Toulmin's view is similar as we have already noticed (cf. I). He thinks men who accept different ideals and paradigms see different phenomena ([83], p. 5); theories determine what are facts for us at all ([83], p. 95). Kuhn too believes that after a revolution scientists work in a different observational world; what they experience has changed ([42], pp. 110, 117, 119, 134, 149). The general problem is the following: Assume we accept a scientific theory, T_1. What we experience, E_1, will now be different from what we would experience if we accepted any significantly different "alternative" theory. How, therefore, could we, in a rational way, ever significantly *revise* our theories and beliefs so that they would be about what we would still experience? On their view the moment we changed our beliefs regarding the essential properties of experience, experience itself would change. In short, it would be impossible for us to revise our essential beliefs about what we would still be able to experience. The second we revise them they would no longer be about what we could still experience. This would imply that a scientist's revised essential beliefs about experience are at any time not about what he could experience at that time (cf. pp. 9–11). And this is absurd. If these beliefs were scientific then they would be confirmable or falsifiable and would purport to express truths about what we could experience. But if there were nothing that we *could* experience which would satisfy or falsify these beliefs at the time we hold them (in virtue of our very holding of them) then, indeed, these beliefs would be neither confirmable nor falsifiable for us. Thus, in particular, they would not be *scientific* and would not express truths about what we could experience.

B. So much for the first point. Now for the second, which is quite similar. Feyerabend's, Hanson's, Toulmin's, and Kuhn's views, I maintain, prevent the scientific theory after a revolution, T_2, from being a *rival*, or *alternative*, to the scientific theory, T_1, before the revolution. On their view T_1 determines experience E_1 which is different than E_2, the experience determined by T_2. But given this it becomes difficult, if not impossible, for the scientist accepting T_1 to have professional disagreement about the experience E_1 with the scientist accepting T_2; this is because they are not talking about the same experience or world; in the sense relevant to science it is claimed that they don't experience the same things; $E_1 \neq E_2$. And each scientist is talking about what he experiences. Their beliefs about experience and the world are not, therefore, *rival* beliefs. Nor are T_1 and T_2 *alternatives*. What is T_1 an *alternative to*? It is not an alternative to T_2 for on their view T_2 is talking about something different (E_2) than T_1 is (E_1). T_1 and T_2 are not alternative views of the same world. On their view the world has radically changed. Nor are they alternative views of experience, for this is radically different. If I accept a current version of quantum mechanics and you accept a current version of sociology, we are not thereby disagreeing. We are not thereby expressing rival or alternative beliefs. Just so with any T_1 and T_2 on the above view. If I observe a duck and claim it can fly while you observe a rabbit and claim it can't fly, we are not expressing rival or alternative beliefs. We are seeing different things; thus, our respective claims about them are not *rival* or *alternative* claims. Just so, if one scientist experiences different things than another, then the beliefs they express about their objects of experience are not rival or alternative beliefs.[10]

This consequence is further directly opposed to one of the principal methodological considerations that motivates Feyerabend to hold the radical observational variance position in the first place, namely, the "principle of proliferation: Invent, and elaborate theories which are *inconsistent* (my italics) with the accepted point of view, even if the latter should happen to be highly confirmed and generally accepted" ([18], pp. 223–224). This principle is central to Feyerabend's positive methodology and is a "main consequence" of his "abstract model for the acquisition of knowledge" ([18], p. 223). This principle, under the name of 'theoretical pluralism', "is assumed to be an *essential feature* (his italics) of all knowledge", for "Criticism must use alternatives" ([15], pp. 6–7). Yet it is precisely his radical observational variance position which, as we have seen, precludes two different theories from contradicting each other and from being inconsistent. Rather than being "an *essential* feature of all knowledge", the principle of proliferation would instead be an *impossible feature* of all knowledge. And on a normative interpretation it would, in effect, call for the impossible given the doctrine of radical observational variance. Prescriptions for the impossible, though not strictly contradictory, are somewhat awkward and become especially suspect employed within a *scientific* methodology. Given all this I therefore find it quite ironic that Feyerabend puts forward precisely this methodological principle as a *reason* for his espousal of radical observational variance (cf., e.g. [15], pp. 7–8).

C. A third problem which besets their views is one concerning the range of fact, and of interaction between theory and fact.

Kuhn, for example, denies that during a scientific revolution there exist fixed data which the scientist interprets ([42], p. 121). Yet he also claims that the scientists' world is determined "jointly by the environment and the particular normal-scientific tradition" ([42], p. 111, cf. also p. 122). Here Kuhn presumes that the "environment" is itself unaffected by theory; it is rather the "world" which is so affected. The environment combines with the theory to form the world. In this sense the environment may itself be said to be "fixed". Kuhn gives no reasons why scientists should not study the nature and properties of this environment. Indeed its investigation would be empirical and would be, for this reason, worthy of scientists' attention. Yet it is presumed fixed and unaffected by theory. It would therefore be neutrally available to scientists even during scientific revolutions when different theories are supposedly combining with it to form different worlds. But it is just the availability of fixed data during scientific revolutions which Kuhn, as we have noted, would deny.

There is, in addition, an interaction problem with the radical observational variance view. Several important and related questions naturally arise given this view; and *prima facie*, at least, they are legitimate questions. *How* does the "normal-scientific tradition" operate on, and change, this environment to form the scientists' world? Just how do paradigms or theories interact with the "common environment" to produce "data" ([42], pp. 122, 134; [83], pp. 95–96)? What is the nature of this "manufacturing process" ([14], pp. 50–51)? What happens when new data arise? Where is the *locus* of this interaction? If "However construed, the construing is there in the seeing" ([27], p. 23) then we may ask, precisely *where* in the seeing is it. It does not help whatsoever to say that what happens is a "sudden and unstructured event" ([42], p. 121). This "answer" is nothing more than an expression of a basic unintelligibility of their view.

More traditional approaches to the philosophy of science would have scientific activity consist in the interpretation of neutral and fixed observational data. And a virtue of such approaches is that these questions simply do not arise; *prima facie*, at least, they would not be legitimate questions within the conceptual framework of a more traditional philosophy of science. The problem is that answers to such questions are not explicitly provided by either Hanson, Kuhn, Toulmin, or Feyerabend. Nor do the answers, in an obvious way, follow from, or inhere implicitly in, the conceptual scheme they employ to analyze science; yet it is this very scheme which gives rise to such questions. In short, these writers, in my opinion, lack a well-thought-out and developed epistemology. Is there any reason why they cannot use some accepted epistemological theory to deal with, and perhaps answer, these questions? This is, of course, an open and a difficult question. Many different general epistemological theories have made their appearance in the history of thought and the approach of Feyerabend, Kuhn, Hanson, and Toulmin is perhaps not incompatible with all of them. Their approach is, however, incompatible with either a Platonistic or Kantian epistemology. Consider a Platonistic approach. Here objects of experience and observation would be constituted through participation in *invariant* Forms which exist whether or not we believe that they exist. Such objects would thus be constituted in a way which is independent of our theoretical *beliefs*. Platonistic answers to the above questions would therefore be inappropri-

ate for they would invoke a theory which is incompatible with the radical
observational variance view. On a Kantian approach objects of experience
would be constituted by our Categories and Forms of Intuition each of
which is essentially invariant from person to person. Objects of experience
for Kant are objective, invariant, and theory-neutral because they are
constituted, not by particular scientific theories, but by general and invariant Categories and Forms of Intuition. Such objects would therefore
be constituted in a way which is independent of our particular scientific
beliefs. They would also be independent of our general philosophical
beliefs as to (say) *which* Categories and Forms of Intuition experience must
presuppose; one of the purposes Kant had in writing the *Critique of Pure
Reason* was to convince his disbelieving philosophical opponents that
their experiential world already depended on Kantian Categories and
Forms of Intuition – whether or not these opponents initially believed
that Kantian Categories and Forms of Intuition were correct. Kantian
answers to the above questions would therefore be inappropriate. They
would invoke an epistemological theory which is incompatible with the
radical observational variance view.

D. The fourth problem which besets their views is that if they were true
it would follow that no theory could be tested or falsified by any observations. According to Hanson, Feyerabend, Kuhn, and Toulmin, observations presuppose, and are laden with, the particular scientific tradition of
the time. Therefore, I will argue that observations and observation reports
could not lead to the rational rejection of the scientific tradition. Nor
could they lead to the rational acceptance of a new and revolutionary
tradition.

Hanson, unlike Kant, indeed feels that what is observed presupposes,
and is laden with the particular theory of the time ([27], pp. 10, 23, 149,
157; [28], p. 38):

[the uncertainty principle] is built into the outlook of the quantum physicist, into
every observation of every fruitful experiment since 1925. The facts recorded in the
last thirty years of physics are unintelligible except against this conceptual backdrop.
([27], p. 149)
The observations and the experiments are infused with the concepts; they are loaded
with the theories. ([27], p. 157)

Feyerabend also feels that observational results cannot be arrived at, stated, or verified independently of the particular theories investigated ([16], pp. 151, 194–197, 214, 220–221).

Kuhn ([42], pp. 6, 110–111, 117, 119, 128, 134, 149) and Toulmin ([83], pp. 57, 95, 101, 103; [85], p. 463) concur. Indeed, these views may be part of what prompts Shapere to remark:

The view that, fundamental to scientific investigation and development, there are certain very pervasive sorts of *presuppositions* (my italics), is the chief substantive characteristic of what I have called the new revolution in the philosophy of science... ([78], p. 48)

Such a view on their part is, in my opinion, implausible for two reasons.

1. Consider the first. Let us assume that observations made within a particular scientific tradition presuppose, and are "loaded with", the scientific theory T employed at that time. Any observation report O of these observations will then presuppose T. On Van Fraassen's account of the concept of presupposition ([86], esp. pp. 137–138), the only developed account of this notion known to me, it then follows that O is neither true nor false unless T is true. And from this it follows that if O is true then T is true and if O is false then T is true. Thus, we cannot maintain that O is true and T is false; nor can we maintain that O is false and T is false. The latter result is rather remarkable; it implies that false predictions cannot be used as a basis for denying T. The situation, however, is still worse. No observation report O, whether true or false, can be used to reject T; if O presupposes T then, whenever O is true, T is true, and whenever O is false, T is true. The only way out of this dilemma is to maintain that reports of observations are neither true nor false. But this is tantamount to denying that science is an empirical and cognitive enterprise. In any case, we would have to regard as neither true nor false *all* reports of observations made during former times, if we regarded as false the theory employed during those times. In particular, the observation reports of Brahe, Kepler, Newton, Leverrier, Priestley, Gibbs, and Michelson-Morley would have to be regarded as neither true nor false.

2. Consider the second point. No true or false observation report which presupposes a theory T could serve as a basis for the rational acceptance of a new theory T^1 which is mutually inconsistent with T. If we accept T^1, we would have to deny T, given T and T^1 are mutually inconsistent. If we deny T we would have to maintain that O is neither true nor false. If we maintain that O is neither true nor false it is hard to

see how O could serve as a *basis* for the rational acceptance of T^1. In particular, if we accept relativistic physics (T^1) we would have to deny classical physics (T) since these two theories are mutually inconsistent. If we deny classical physics then we would have to maintain that the observation reports (O) of the Michelson-Morley experiment, of Leverrier's observations of Mercury's rotation round the sun, etc., were neither true nor false. Being neither true nor false it is hard to see how they could have served as a basis for the rational acceptance of relativistic physics over classical physics. These observation reports, however, *are* generally considered by scientists to be correct and true, and are part of what led to the rational acceptance of relativistic physics over classical physics (cf., e.g. [54], pp. 542–547).

VI

We have, therefore, found some methodological justification for assuming that, in a non-trivial sense, experience is neutral with respect to alternative scientific theories. At best the contradictory view implies that theory change is not rational and without justification. At worst, the contradictory view leads to unintelligible and absurd consequences.

In addition to such methodological justification, the historical examples that we discussed (esp. in IV) have suggested that scientific observations are neutral. There are two bits of further evidence for this conclusion.

The first is the existence of phenomena such as surprise and unsettlement which occur during scientific transitions. Scheffler notes this well:

Observational support for an assignment contrary to an accepted hypothesis needs to persist longer and fight harder for a hearing than observational data which accord with expectations... Yet such contrary indications can and do make themselves felt. Our expectations strongly structure what we see, but do not wholly eliminate unexpected sights. To suppose that they do would be, absurdly, to deny the common phenomena of surprise, shock, and astonishment, as well as the reorientation of belief consequent upon them. ([73], p. 44)

The second bit of evidence is that scientists in different traditions sometimes use the same sorts of sentences to describe what they observe. This approach has been used by Feyerabend to suggest the opposite conclusion, that of "the dependence of perception on belief" ([16], p. 220). He calls to our attention "the existence of *genuine observational reports* concerning devils and gods" ([16], p. 220, his italics). According to Feyerabend,

Numerous eyewitnesses claim that they have actually *seen* the devil, or *experienced* demonic influence. There is no reason to suspect that they were lying. Nor is there any reason to assume that they were sloppy observers. ([15], p. 32)

Therefore, "...primitive people, whose life is governed by a powerful myth, live in an observational world very different from our own..." ([16], pp. 220–221). For Feyerabend this is presumably because in order to describe what they observed they sometimes used different sorts of sentences – sentences from a different conceptual framework – from those we would use.

There are three objections one might raise in connection with Feyerabend's remarks here. First, we could claim that many of the previous scientific observations made by scientists in different traditions are, or were, experimentally *repeatable*, whereas those Feyerabend mentions are not. Second, his evidence is incompatible with his doctrine of radical meaning variance (a doctrine we will discuss in detail in Chapters 2 and 3). Consider the following statement:

(19)   I saw a devil.

Feyerabend uses the fact that people living in a different tradition than ours have uttered (19) in order to suggest that what they observed, viz. devils, was radically different than what we observe (cf. [15], p. 32). However, if 'devil' means something radically different for them than it does for us, as his doctrine of radical meaning variance would imply, we cannot use (19) as evidence that they observed something radically different than what we observe; we cannot use (19) as evidence that they observed what *we* would call 'a devil'. If, on the other hand, 'devil' is held to mean the same for them as it does for us, then Feyerabend's argument presupposes meaning-invariance. The third objection is that even if Feyerabend's argument is correct all that follows is that observational data *sometimes* are not neutral and *sometimes* depend upon belief, not that they *always* do, or that they must. Even, if correct, the conclusion of his argument is not at all opposed to (say) Scheffler's remarks quoted above, which stressed the possibility of neutral observations.

VII

There are, then, difficulties in the positions of Feyerabend, Hanson, Kuhn, and Toulmin as to observation in science. I have tried to show that an alternative to their view is preferable. Nevertheless, their position does suggest something valuable. Scientific revolutions did not consist merely in the discovery of new facts or merely in a closer attention to facts already known. Not everything a scientist observes has equal potential for testing or confirming his theory. Some things a scientist experiences may be relatively irrelevant to his theory. In this sense we may say that different theories define different realms of experience; that is, the potential of the experiences for confirmation and test have changed with change of theory. In this sense we may say that observation is theory-laden. Some observations change in importance when theories change. An observation, or type of observation, that used to be important with respect to a theory, T_1, may no longer be as important when we abandon T_1 and accept, and use, another theory, T_2. And this entails a sounder view of the history of science. Past scientists, as Feyerabend, Hanson, Kuhn, and Toulmin recognize well, need not be blamed for lack of attention to observational detail which we attend to. They need not be blamed for overlooking things we do not overlook; this is because they were sometimes looking for different things – the things that would (a) resolve the particular problems and anomalies engendered by *their* theory and (b) confirm and test their theory to a high degree [(a) and (b) are not unrelated]. Past scientists need not be blamed for not performing the experiments that present scientists perform; and this even if they did have the experimental apparatus which would have enabled them to do so. As Toulmin aptly puts it "...the questions we ask [nature] inevitably depend on prior theoretical considerations" ([83], p. 101). And different theories may ask different questions, yield types of predictions (some more central than others) with differing logical interconnections in need of different types of observational test. Hanson is also sensitive to this point. He suggests that differences in their conceptual organization more or less guaranteed that Mach and Hertz would not have the same research problems, would not perform the same experiments ([27], p. 118). The *direction* of each's inquiry was somewhat different; and this was most likely due to a differing internal and logical, or conceptual structure between the two theories. Tentative theories or hypotheses are needed to give direction to a scientific investigation. They determine what data should be collected at a given point in a scientific investigation. Data are highly relevant to a tentative theory which is advanced for consideration when either their occurrence or non-occurrence can in some sense be centrally inferred from this theory. And only these data – the relevant data – should be collected. Some thinkers have unfortunately denied this. As Feyerabend puts it,

What *is* to be criticized is the attempt *without* much guidance from thought to collect as many useless facts as possible from as many domains as possible, and to expect that science will one day miraculously profit from the collection thus assembled. ([16], p. 156)

Past scientists did, in fact, perform different experiments from present scientists. Moreover, they should have. They were testing a different theory than present-day scientists are testing. What a scientist ends up seeing and observing partly depends on what he is looking for. Scientists who are looking for new things obviously end up observing new things, if they find what they are looking for. (Note that this does not imply that they could not still see the old things if they looked.) In this sense, also, observation is "theory-laden". But, as we argued earlier in this Chapter, any stronger sense is implausible. Nevertheless, the sense in which observation *is* theory-laden is not an uninteresting sense. Different theories highlight different features of the world. The scientist's attention is directed to those parts of experience which, in Kuhn's words, "promise opportunity for the fruitful elaboration of an accepted paradigm" ([42], p. 125). Scientists do, and should, devote little attention to processes which are, in Feyerabend's words, "situated at the periphery of their enterprise" ([16], p. 152). "Science", as Kuhn claims, indeed "does not deal in all possible laboratory manipulations" ([42], p. 125). It selects only those relevant for testing a theory currently under consideration. "As a result, scientists with different paradigms engage in different concrete laboratory manipulations." "After a scientific revolution many old measurements and manipulations become irrelevant and are replaced by others instead" ([42], pp. 125, 128). There is nothing pernicious about the theory-ladenness of observations in *this* sense. After a revolution, theory T_2 still accounts for the observations and phenomena that T_1, the previous theory, accounted for. But T_2 does more. It accounts for what T_1 didn't account for. T_2 also generates some new predictions. And keeping in mind that T_2 accounts for the phenomena that T_1 accounted for, scientists then direct their attention to test the outcome of these new predictions. They may, in carrying out these new tests and experiments, observe some new types of experience. But this does not imply they no longer see what they saw before. They still see the old things. But they don't bother with them after their occurrence has been accounted for by the new theory. Their attention becomes refocused. They look for additional things. They see and observe additional things. The realm of experience has *widened* with the acceptance of T_2 to include these new observations.[11] But this does not imply that previous observations, made using T_1, are somehow either annihilated, changed, or no longer accounted for by T_2. Widening is not annihilating. Explaining is not explaining away.

Return to an earlier example for a moment (IIC). Mohammed X, the black Muslim, possesses a quite different set of beliefs than Adolph Übermensch, the Nazi. Each would interrogate the young heavyweight (Cassius Clay = Mohammed Ali) somewhat differently. Each would ask him somewhat different questions before deciding whether he should be stripped of his championship and imprisoned. And they might differ as to his guilt or innocence. Yet, in spite of their widely differing beliefs they would have asked him many of the same questions. Enough of these would ensure that they were each talking about the same man. Namely, each would inquire as to his place of birth, the number of his professional fights, his height, his weight, etc. Just so for scientific theories. The fact that scientists ask different questions of nature does not imply that they are not talking about the same things. It does not imply that there is not a large overlap between their two realms of experience; it does not imply one of the realms is not an extension of the other.

CONCLUSION

Recognizing something valuable in Feyerabend's, Hanson's, Kuhn's and Toulmin's position on observation, we must still reject much of it for the reasons given throughout this Chapter; their whole position taken as is destroys the possibility of comparing and judging theories by reference to experience. We have urged that the use of Gestalt examples as the paradigmatic case for science tends to generate more problems than it solves. Hanson urges that after a student has received scientific training he no longer sees the same "glass and metal instrument" ([27], p. 15) he saw before. He now "sees the instrument in terms of electrical circuit theory, thermodynamic theory,... [etc.]" ([27], p. 15). We have urged, instead, that such statements mean nothing other than that the student now has new *beliefs*, perhaps expressed by means of new concepts, about the instrument, about what it does and is, and about how it works. We have urged that the claim that such a scientist *sees* something different is unjustified. Believing is not seeing.

In fact, it has not been shown that facts are not neutral. We have suggested that scientific facts are neutral. Moreover, we have suggested that it is methodologically desirable that they be so: for the purposes of being able to compare, judge, and revise, alternative and rival theories by reference to experience; for the purpose of there being able to *be* alternative and rival theories at all; for the purpose of avoiding an interaction problem between theory and environment; for the purpose of being able to test or falsify a theory by reference to observations. And none of what we have maintained is inconsistent with our claim that different theories redi ect and refocus the scientist's attention or with our claim that different theories lead the scientist to *look for* different things. New theories, if they are better theories, *do, and should,* cause us to examine and observe, new and additional aspects of sensory experience; they should cause us to attend to more and more of this experience. In this sense, we may say that what scientists observe does change; it increases. In this sense, we may say that observation is theory-laden. This sense is not pernicious, nor is it inconsistent with what we have maintained above; what a scientist observes, and therefore sees, before he accepts a new theory, is still the same; it remains recorded; and he could see it again if he looked. A new theory usually *does*, and moreover *should*, account for previous observations made when using a previous, *rival* theory. But the new theory *also* accounts for more. It accounts for observations the previous theory positively failed to account for. *And* it accounts for observations the older theory didn't even talk about. In this sense experience has widened. New theories are more *comprehensive* than old theories. Science is, therefore, a cumulative and expanding enterprise.

NOTES

1 Some might here object that this should preclude Kuhn from describing the *structure* of scientific revolutions except in the sense that they have no structure. I do not feel that such an objection is entirely forceless.

2 Not only do Tycho and Kepler see different things, but laymen are compared to

infants and idiots in that they are held to all be literally blind to what the physicist sees ([27], pp. 17, 20, 22).

3 Hanson implicitly claims "seeing X as Y" is the same as "seeing an object X". Namely, seeing X as Y *is to* see that if $A_1, \ldots A_n$ were done to X then $B_1, \ldots B_n$ respectively would result (cf. [27], p. 21, where Hanson discusses Figure 1). This is, unfortunately, incompatible with Hanson's explicit claim that he is not identifying *seeing* with *seeing-as*: "I do not mean to identify seeing with seeing as" ([27], p. 19). Since 'sees that' has the same sense in both claims, so would Hanson's two types of "seeing".

4 All of this, in effect, *assumes* that some of the intersections of some of the respective categories employed by different scientific schemes are non-empty. It is unfortunate that Scheffler does not analyze Feyerabend's denial of precisely this assumption [17]. We will try to remedy this in Chapter 2, Section IV and again in Chapter 4 when we return to the topic of observational invariance.

5 Toulmin notes this well, ([84], pp. 83–85) and interprets Kuhn as very recently emending his "normal-revolutionary" distinction to one in which theoretical *micro-* revolutions are going on continuously, [43]. Purtill is excellent on this point [62].

6 This point will be elaborated upon in Chapters 2 and 3 and some much-discussed examples will be rediscussed.

7 Cf. "After investigating the matter further, I now agree with you. I did not, as I had previously thought, see John commit the crime at all. As you say, John definitely was in Japan at the time. It was, after all, Charles whom I saw."

8 It is my opinion that many of Kuhn's claims in [42] are either false or else provide additional examples of what Professor Ryle has termed "systematically misleading expressions" in his now classic article, [69]. Cf.: "When an expression is of such a syntactical form that it is improper to the fact recorded, it is systematically misleading in that it naturally suggests to some people – though not to "ordinary" people – that the state of affairs recorded is quite a different sort of state of affairs from that which it in fact is." ([69], p. 16)

9 They will be "efficient", according to Feyerabend, if they fulfill in good measure the methodological aim of mutual criticism. (Cf. [15], pp. 7–8)

10 Perhaps there may be forms of cognitive conflict other than contradiction. Even if this may be so – although it is not at all clear to me how it could – my above argument does hold for a large class of conflicts; it shows that cognitive conflict in the sense of contradiction would be precluded. I will examine, in Chapter 3 (IIB), two special senses of conflict, other than contradiction, which Feyerabend has proposed.

11 By suggesting the unification of terrestrial and celestial laws, the Copernican revolution made the projectile a legitimate source of information about planetary motions. (Cf. [41], p. 230)

CHAPTER 2

AN EXAMINATION OF SOME ARGUMENTS AND CRITERIA FOR RADICAL MEANING VARIANCE

In Chapter One I argued that it has not been demonstrated that observation is not neutral. I have suggested that in important and typical cases observation is neutral and critically independent of particular scientific hypotheses. I noted that one need not introduce a dubious notion of the given in order to make such independence possible. Further, I argued that the neutrality of observation is methodologically desirable.[1]

I noted that the historical examples we examined have suggested that scientific facts are neutral. Some further evidence for this conclusion was that scientists in different traditions sometimes use the same sorts of sentences to describe what they observe.

The second bit of evidence would be objected to by those who advocate radical meaning variance. They would claim that although scientists within different traditions accept some observation statements using the same terms, these terms are radically different in meaning. For this reason the observation statements in which such terms occur are held to be radically different in meaning. Thus, they would conclude, using the same sentences is not evidence that the same things are being described (cf., e.g. [27], pp. 7, 23). Devising an answer to such an objection forces us to also deal with the general arguments and criteria adduced in support of radical meaning variance. These arguments and criteria, however, are important in their own right and central to current controversy in the philosophy of science. They therefore warrant close attention and critical evaluation. There is much similarity here in the views of Feyerabend, Hanson, Hesse, Kuhn, and Toulmin. Each is sympathetic to, and adopts, a radical meaning (as well as observational) variance position with regard to most scientific transitions. There are differences, of course, among their views, and each in some respects differs from the others. However, it is not the differences, but rather the similarities that will concern us here. We will soon note these similarities. Because of them we shall refer to these philosophers of science as radical meaning variance theorists.

I

The radical meaning variance position usually begins by noting that terms do not possess meanings in virtue of an isolated attachment to their sound or typographic shape; rather, it is held that terms possess meanings in virtue of their systematic function within a theoretical framework or system. The meaning of a term is dependent upon the theory in which it has a place; to change this theory, it is claimed, is to alter the term's relative location with respect to other terms and therefore its meaning. To adopt a new theory is to reassign the roles of both theoretical and observational terms. Such meaning change is, on their view, *radical*. That is, it would preclude important *comparisons* of different theories through appeal to some sort of shared meaning of the terms employed. The comparisons here would include whether or not one theory *is mutually incompatible with, disagrees with, is a rival of, is an alternative of, is in competition with, is reducible to, is derivable from, is better than,* or *is more acceptable than,* another theory. This alone is, of course, not to say that radical meaning variance theorists would have to or want to deny the possibility of these comparisons altogether. Indeed, they would want to, and do try to, make these comparisons between different theories (cf. Chapter 3). The point is that, on their view, these comparisons are not possible through appeal to the meanings of the terms employed. There would not, on their view, be enough shared meaning between the terms occurring in different theories to serve as a basis for these comparisons; they would maintain that the terms express incommensurable concepts.

The following two central theses emerge from such a view:

(1)   The meaning of any scientific term depends on the theoretical context in which it occurs.[2]

(2)   The meaning of any scientific term which occurs in a theory will change radically if that theory is modified.[3]

I have no wish to contest thesis (1) which stresses some sort of dependence of meanings upon theoretical or linguistic context. Rather, I think it plausible. Surely, not all meanings are inherent in the physical or typographic constitution of terms. The same term, considered typographically merely as a physical pattern, *may* have different meaning in different circumstances. For example, it may undergo historical changes in meaning, take on different nuances in different practical contexts, occur with different import in different languages, change with the speaker, or be subjected to various technical stipulations. The important factor, then, seems not to be the typographic or physical character of the term; rather, it at least sometimes seems to be the role given to it by its user in the human context of its use. And this seems true for *both* the sense and reference of terms; neither need be inherent in the term's physical or typographic constitution.

I now want to argue, however, that accepting thesis (1) does not, as is frequently intimated by radical meaning variance theorists, further compel one to accept theses (2).[4] Meaning is not, as I have just noted, a function only of the typographical constitution of terms. Meanings indeed may be regarded as internal or relative to a given theory. But this does not compel us to assent to the further belief that they cannot be shared by other theories. From the mere fact that meanings depend on theoretical contexts it does not at all follow that they therefore bear a one-to-one relationship to theoretical contexts. Not every relationship is a one-to-one relationship. Thus, (2) does not follow from (1).

I do feel that (1) is plausible in the sense specified. Let us therefore inquire into (2) and examine the reasons other than (1) which either perhaps could be, or in fact are, adduced in support of it.

It might prove illuminating to first approach this question by way of some general philosophical positions concerning the theory of meaning. Usually "the problem of meaning" is taken to have two aspects, namely, significance or the having of meaning, and likeness of meaning or synonymy. The latter aspect of the problem – likeness of meaning or synonymy – is *prima facie* involved in claiming (2), viz., that transitions from one scientific theory to another force radical changes in the meanings of the terms employed.

Widely varied approaches have been put forward in attempting to find out under what circumstances two terms, from either the same or different languages, have the same meaning. One of the earliest approaches is that two terms have the same meaning if and only if they stand for the same Platonic Form or Real Essence. Another proposal is that two terms have the same meaning if and only if they stand for the same mental image or picture. A third approach is that two terms have the same meaning if and only if we cannot conceive – not merely cannot imagine or picture –

something that satisfies one but not the other. The third approach is taken to be more general than the second, for presumably we can conceive of a seven-dimensional figure but not imagine or picture it. A fourth criterion is that two terms have the same meaning if and only if there is nothing possible that satisfies one but not the other. A fifth proposal is nominalistic: two terms have the same meaning if and only if they have the same extension.

The latter are traditional ways one might attempt to demonstrate meaning change. Only the first (cf. II), fourth (cf. II), and fifth (cf. III and IV) seem to be involved in, presupposed by, or motivate, contemporary arguments for radical meaning variance. This is perhaps because of the difficulties such approaches have encountered.[5]

Some more recent approaches concerning synonymy and likeness of meaning have been taken by Alston, Mates, Quine, and Goodman.

Alston ([4], p. 37), following Austin's lead ([6], lecture viii, ff.), claims that two words have the same meaning if and only if they "can be substituted for each other in a wide range of sentences without altering the illocutionary act potentials of those sentences." This criterion has, however, not been used by radical meaning variance theorists to establish their conclusions. The applicability of Alston's criterion to actual scientific transitions would be at best problematic; this is due to difficulties with the notion of illocutionary act potential along with difficulties in deciding whether two scientific sentences *in fact* have the same illocutionary act potential.

It is unclear whether radical meaning variance theorists regard scientific change as change of language. If they do then Mates' criterion would also be of dubious help, for it is restricted to *intra*-language synonymy:

Two expressions are synonymous in a language L if and only if they may be interchanged in each sentence in L without altering the truth value of that sentence. ([55], p. 119)

It is doubtful whether this would work as a criterion for synonymy across languages as Mates himself notes.[6] Further, for reasons similar to those that I will present in III and IV, a diagnosis of meaning change using his criterion would presuppose that there has been some meaning invariance; it could not therefore support the radical meaning variance thesis, (2).

Quine and Goodman each express doubts concerning synonymy with respect to both intra- and inter-language translation. Each stresses some sort of nominalistic notion of extension as being the important matter. And each would conclude that no two terms are exactly synonymous; rather, they would opt for a notion of synonymy which would be one of degree or likeness of meaning ([64], p. 63; [24], pp. 73–74). Radical meaning variance theorists could be construed as taking seriously such doubts about the possibility of exact synonymy. And more importantly they could be construed as extending these doubts to cover the possibility of two terms being commensurable, similar, or alike in meaning to *any* degree when such terms occur in different theories.[7] For this they could not rely merely on Quine's and Goodman's conclusion that no two terms are exactly synonymous; such a conclusion is not sufficient to establish the radical meaning variance thesis which maintains that there is a radical and incommensurable difference in meaning between terms occurring in different scientific theories. Indeed, Goodman's and Quine's conclusion is compatible with the claim that, for purposes of mutual comparison, there is a sufficient degree of meaning invariance involved between terms occurring in different scientific theories.

Specific arguments are therefore needed in order to establish the radical meaning variance claim, for this claim does not follow in an obvious way from most of the customary philosophical positions concerning the theory of meaning.

II

To these arguments we will now turn. They seem to be of three varieties.
The first is typified by the following outline:[8]

(3)   The meaning of any scientific term S changes radically if it enters into different essential relations with other terms.

(4)   Any scientific term S which occurs in a theory T enters into different essential relations with other terms if T is modified.

(5)   Therefore, if T is modified then the meaning of any scientific term which occurs in it has changed radically.

The conclusion amounts to the claim that *any* modification of T entails that the meaning of *each* term radically changes.

The inference from (3) and (4) to (5) is indeed valid. Both (3) and (4) are, however, problematic.

A. Consider (4). Perhaps any modification making T a different *theory* will make essential changes. Nevertheless, it does not follow that any modification of T can reasonably be said to make it a different theory. There exist abundant examples of modifications of theories which clearly are too minor to warrant the label 'change of theory'. A few would be as follows: the addition of an epicycle; a change in the value of a constant. And such minor modifications which cannot be said to make T a different theory also do not force the terms occurring within T to enter into different essential relations. (4) is therefore false. If *any* modification of T forced each term to enter into different essential relations then we would have to abandon the assumed distinction between essential and non-essential relations; and if we do abandon such a distinction the above argument collapses for it requires that distinction by having (3) and (4) as premises.

An emendation of (4) which meets this objection is, however, readily available, namely (4'):

> (4')   Any scientific term S which occurs in a theory T enters into different essential relations with other terms if T is modified so as to make it a different theory.

There is, however, a problem of how we are to apply (4') to actual scientific transitions. It would be difficult, given a radical contextualist position, to decide in a non-arbitrary way which modifications of a theory make it different.

B. More serious problems, however, arise with (3). *Which* of the essential relations must change for S to change radically in meaning? (a) Must every essential relation that S enters into change for the meaning of S to change radically? (b) Or need only *some* change?

1. If the answer to (a) is in the affirmative then the application of (3) by meaning-variance theorists to actual scientific transitions is clearly incorrect. For not every essential relation has changed for the terms in their examples. Consider the much discussed example of 'mass' ([14], pp. 80–81; [15], pp. 14–15; [16], pp. 168–170; [42], p. 101). In *both* Newtonian mechanics and relativistic mechanics, 'mass' enters into some unchanged essential relations with "other terms", namely, mass can be characterized in both theories as the ratio of force to acceleration; thus, 'mass' enters into the same relation, "equals", with 'F/a' in both theories.

2. On the other hand, assume that we answer in the affirmative to (b). (3), on this interpretation, becomes (3'):

(3')   The meaning of any scientific term changes radically if it enters into some different essential relations with other terms.

(3'), however, is false. If a scientific term enters into *some* different essential relations with other terms then what follows, I think, is that the term has acquired *a* new meaning, The term may, however, also retain some of the essential relations that it enters into with other terms. Platonistically speaking, we might therefore say that part of its old meaning is retained. That the term has some additional uses and some different uses does not imply that there are not some uses which remain the same. Speaking from the viewpoint of Wittgenstein (or Alston or Austin) we might therefore say that part of its old meaning is retained. That the term refers to some new entities and some different entities does not imply that there are not some things which it refers to which are the same in both theories; that the extension of the term is different in each theory or framework does not imply that the intersection of these respective extensions is trivial or null. Speaking from the viewpoint of Quine, Goodman, or Scheffler, we might therefore say that part of its old meaning is retained with respect to revisions of theories or frameworks. In *either* of these three senses of meaning, therefore, it *need not* follow that the modifications are extensive enough to warrant a 'radically changed in meaning' label. It need not follow that the term has acquired completely or radically new meaning. It may be that the meaning of the term has changed, but not radically. Change of meaning need not be "all or none". There are many situations in which asserting,

(6)   *S* has acquired a new meaning

would be sensible, but asserting,

(6')   *S* has acquired radically new meaning

would be rash. If a scientific term enters some different essential relations then although (6) probably will follow, (6') need not. This is not to deny that modifications in a theory may sometimes be of a sort that make it reasonable to assert that the meanings of some of the terms involved have radically or completely changed. The point is simply that although two theories may attribute some different essential properties to items in their common domain,[9] many other essential properties attributed to the items in question may remain the same.

To hold (3′) is to ignore the fact that when a term enters into different essential relations, many of the other essential relations into which it enters may remain constant so as to warrant the claim that the term retains much of the meaning it had before. And *even in those* cases in which an 'unchanged in meaning' label does not seem altogether adequate it need not follow that a 'radically changed in meaning' label is justified. In *such* cases it may be more enlightening simply to note which essential relations relevant for understanding a term have changed, the nature and extent of this change, and which have remained constant, i.e. how, in what respects, and to what degree the term has changed its meaning and how not. What most often follows, I think, when term S in theory T enters into some different essential relations in theory T' is (6). And this is compatible with T' being an extension of T, with T being a limiting case of T', or with T being reducible to T'. Because (6′) is instead held to follow it is just these possibilities that are denied by radical meaning variance theorists.[10]

C. The major difficulties of the argument that we have examined in this section stem from its inability to take cognizance of the various degrees of dependence as well as independence that the significance of terms may exhibit with respect to the theories in which they appear. The root of these difficulties lies in (say) Kuhn's and Feyerabend's rigid conception of what a difference of meaning amounts to when there are framework revisions. They hold, as Shapere ([78], pp. 67–70) and Achinstein ([2], pp. 103–104) have pointed out, that two expressions or terms occurring in different theories or frameworks must either have precisely the same meaning or else it must be radically and completely different. No one should be surprised that the root of the trouble, although not apparent until after some analysis, should turn out to be such a simple point, for philosophical difficulties are often of just this sort. In Chapters 3 and 4 we will attempt to provide a middle ground by altering (say) Kuhn's and Feyerabend's rigid notion of meaning. By claiming that meanings can be similar and comparable in some respects though different in other respects, we will hope to preserve the fact that, e.g. classical and rela-

tivistic mechanics *are* comparable even while being perhaps more fundamentally different than the most usual logical empiricist views make them.
The above source of error, further, will be prominent in subsequent
discussion in this Chapter (cf. IV).

III

There is another, related, line of thought which might be adduced to
support a weaker version of the radical meaning variance thesis. It is
quite similar to the argument of II but we will weaken the conclusion to
claim only that the meanings of the terms change in an ordinary sense,
not that they change radically or completely or "incommensurably" as
(say) Feyerabend and Kuhn would have it. This is no doubt contrary to
the intent of the argument as usually proposed. Any difficulties, however,
that we succeed in raising against our weakened version of the argument
will *a fortiori* hold against the more radical version. Further, we will not
assume a distinction between essential and inessential relations. In the
spirit of Quine (e.g. [65], p. 199) we will assume only a distinction between
important or central relations, and unimportant or non-central ones. The
general strategy is to produce several central or important statements
employing the term in question which hold relative to one theory but fail
to hold relative to the other (cf., e.g. [14], p. 82). For example, the meaning
of 'mass' is held to change in the transition from classical to relativistic
mechanics because it is true that relativistic mass depends on velocity,
whereas classical mass does not; again, relativistic mass is convertible
with energy, whereas classical mass is not, etc. Although one might expect
such examples to demonstrate the mutual inconsistency of the theories,
they are instead offered as evidence that the meaning of the term in
question has changed. We can state the criterion of meaning change at
work here as follows:

(7)   If term S means the same in theories T, T', then each funda-
       mental statement using S and whose other occurrent terms
       remain the same in meaning will hold relative to T' if it holds
       relative to T.

The negation of the consequent of (7) would seem therefore to provide us
with a sufficient condition for the meaning of a term to change.

A. However, the antecedent of (7) requires a consistency condition. This is because if T' and T were mutually inconsistent then even if the meaning of S remained the same, fundamental statements employing S that hold relative to T might not hold relative to T'. That is, if T and T' were mutually inconsistent it could be the case that although S means the same in both, there nevertheless exists a fundamental statement using S and whose other occurrent terms remain the same in meaning which holds relative to T but which does not hold relative to T'. But, in that case, (7) would turn out to be false. Interestingly, this can be illustrated by an example Feyerabend has accepted ([17], p. 267; [18], pp. 230–231). Consider classical celestial mechanics T and a theory T' like classical celestial mechanics except for a slight modification in the strength of the gravitation potential. As Feyerabend admits, T and T' are inconsistent yet there is no change of meaning involved. There also exists at least one fundamental statement (the law of gravitation in T) expressible in both T and T' which holds relative to T but which does not hold relative to T'; and Feyerabend would agree ([17], p. 267) that the terms occurring in this statement have the same meaning in each theory. (7) would therefore be false. To ensure the truth of (7), therefore, it is required that T' be consistent with T. If one adds such a consistency clause to (7), however, the result is merely a sufficient condition for *either* a change in the meaning of S *or* the mutual inconsistency of T and T'.

B. The situation, however, is still worse. The phrase "whose other occurrent terms remain the same in meaning" is necessary in the formulation of (7). To delete it would permit terms other than S to change in meaning; therefore even if the meaning of S were to remain constant, the truth values of statements employing S might well alter. If we include the phrase, however, (7) turns out to require for a "diagnosis" (the term is Feyerabend's) of a change in the meaning of a term exactly what (7) would have to deny, namely, *other terms*, involved in the original fundamental statements, that retain the same meaning from one theory to another. Consider once more the case of 'mass'. The important statements expressing the important connections between mass, velocity, and energy are supposed to provide a basis for the claim that 'mass' changes meaning in the transition from Newton to Einstein. If this is so then, according to (7), the meanings of 'velocity' and 'energy' must be constant. But it is just this that must be denied when in light of these statements we apply (7) to 'energy' and 'velocity' respectively. (7) is, thus, demonstrably inadequate. One must conclude, therefore, that this criterion is not adequate even to support the claim of non-radical meaning change from theory to theory.

IV

In a recent note [17], Feyerabend has modified his earlier views (cf. Note 3) and suggested another criterion of meaning change. In that note, Feyerabend admits that not every change of theory involves change of meaning. He cites as an example a case of two theories T, classical mechanics, and $\bar{T}$, like T except for a slight change in the strength of the gravitation potential. He maintains that T and $\bar{T}$,

> are certainly different theories – in our universe, where no region is free from gravitation influence, no two predictions of T and $\bar{T}$ will coincide. Yet it would be rash to say that the transition $T \rightarrow \bar{T}$ involves a change of meaning. For though the *quantitative values* of the forces differ almost everywhere, there is no reason to assert that this is due to the action of different *kinds of entities*. ([17], p. 267)

Feyerabend thus seems to say that *theories* are different if they assign different *quantitative values* to the magnitude terms employed; on the other hand, the *meanings* of the terms employed are different if they have to do with different *kinds* of entities. He explicitly states his new notion of change of meaning and stability of meaning in the following passage:

> A diagnosis of *stability of meaning* involves two elements. First reference is made to rules according to which objects or events are collected into classes. We may say that such rules determine concepts or kinds of objects. Secondly, it is found that the changes brought about by a new point of view occur *within* the extension of these classes and, therefore, leave the concepts unchanged. Conversely, we shall diagnose a *change of meaning* either if a new theory entails that all concepts of the preceding theory have extension zero or if it introduces rules which cannot be interpreted as attributing specific properties to objects within already existing classes, but which change the system of classes itself. ([17], p. 268)

A. Let us evaluate this analysis. It depends on being able to collect objects into classes; and, as Shapere notes ([78], p. 63), "this in turn rests on being able to refer to "rules" for so collecting them". Such rules determine the classes and presumably can change the system of classes. If the changes occur only *within* the extensions of these classes the meanings will not have changed. But, if the new theory changes the whole system of classes (or "entails that all concepts of the preceding theory have extension zero") then the meanings will have changed. In order "to apply this criterion, the rules of classification must be unique and determinate"; they must allow "an unambiguous classification" of the objects involved ([78], p. 64). Otherwise, we might not be able to determine whether the change affected the entire system of classes, or whether it merely occurred *within* the extension of the previous classes. If so, we would not be able to determine whether or not, on Feyerabend's criterion, a change of meaning had in fact taken place. As Shapere has pointed out ([78], p. 64) this is, indeed, most often the case; there may well be two different sets of rules and consequent systems of classification, and only one may imply a change of meaning has taken place. One can, in scientific usage, collect objects into classes in a variety of ways, and on the basis of a variety of considerations ("rules") and purposes. And the scheme of classification we choose will depend largely on the context and on our purposes, not simply on intrinsic properties of the objects involved.[11] Are neutrons different "kinds of entities" from electrons and protons? Or are they simply a different subclass of elementary particles? Are gold objects different "kinds of entities" from "silver objects"? Or are they simply a different subclass of metals? Are the light rays of classical mechanics and of general relativity (two theories which Feyerabend claims are "incommensurable") different kinds of entities or not? Such questions can be answered *either* way, depending on the context and on our purposes. For example, one could say that neutrons are a different kind of entity from electrons and protons, for the former carry no charge, whereas the latter do. Yet, it is also true that neutrons, protons, and electrons are each elementary particles and in this sense are, thus, not different kinds of entities. Given Feyerabend's criterion it is just not clear which of these answers we are supposed to give. This is to say that the original questions, as they stand, are not clear, for there are similarities as well as differences between neutrons and protons, silver and gold, and light rays in classical mechanics and light rays in general relativity. Such questions can be given a simple answer ("different" or "the same") only if unwanted similarities or differences are stipulated away. Feyerabend could, of course, simply claim that each of these entities *is* of a different kind from the other. The problem, however, is how such a claim could be warranted in the light of his above analysis and criterion.

Let us, nevertheless, agree for the moment to Feyerabend's somewhat arbitrary decision "not to pay attention to any *prima facie* similarities that might arise at the observational level, but to base our judgment [as to whether change of meaning has occurred] on the principles only..." ([17], p. 270). Even this need not compel us to assent to his further claim that a radical change occurred in the meanings of the spatiotemporal terms in the transition from classical mechanics (T) to the general theory of relativity ($\bar{T}$). Do the meanings change so much as to preclude the possibility of comparisons (of the sort that we mentioned at the start of this chapter) between the two theories through appeal to shared meaning? Nothing in Feyerabend's new criterion or in the way he employs it [if he *employs* it in his note at all; he really seems to employ something like (7)] would support an affirmative answer. In fact one could argue, as Fine does ([20], pp. 234–235), that the transition is from one metric geometry to another and that the general concept employed is that of a metric (or pseudometric) space. This concept was first introduced by Fréchet ([22], 1906) prior to Einstein's paper on general relativity. The spatiotemporal frameworks of these theories are thus still comparable with respect to their possession of certain kinds of mathematical properties, viz., metric and topological ones; both theories do have something to do with "spaces" in a well–defined mathematical sense. And, as Shapere has pointed out ([78], p. 64), the same question must arise – and is equally useless and at this stage answerable only by arbitrary stipulation – as to whether the spatiotemporal frameworks involved share the same *kinds* of properties and *are* the same *kinds* of entities ("spaces"), or whether these properties are not "specific" ([17], p. 268) enough for us to say that those frameworks involve the same "kinds of entities". It has become common to say that meanings are determined by rules ([58], p. 47; [63], p. 217; [72], p. 144). But from Feyerabend this becomes a useless item of information, for he gives no hint as to how such rules could be extracted from the theories and how they could consistently function to determine meaning.

B. 1. The situation with respect to Feyerabend's new criterion is, however, still worse. It can, in my opinion, be shown to be logically untenable. It would seem to be a criterion not for change in the meaning of a single term but rather for some sort of wholesale change of meaning. This is suggested by one implication of Feyerabend's new criterion:

(8)   "...we shall diagnose a *change of meaning...* if a new theory entails that all concepts of the preceding theory have extension zero..." ([17], p. 268)

(8) is, in my opinion, demonstrably untenable. The possibility of logical relations between the new theory, $\bar{T}$, and the preceding theory, T, such that each concept of T can be taken to be the "same concept" in $\bar{T}$ is what is precisely at issue. Assume $\bar{T}$ entails that all the concepts of T have extension zero. Then, evidently, each concept of T must also be carried over, and common to, $\bar{T}$; if each concept were not also common to $\bar{T}$, then $\bar{T}$ could not "entail" anything about these concepts and in particular $\bar{T}$ could not entail that these concepts have extension zero. Thus, there must be stability of meaning between T and $\bar{T}$ with respect to each concept of T. From (8) and our assumption, however, there must also be a change of meaning for these concepts. Feyerabend's criterion (cf. above) is, thus, logically untenable, for one of its logical implications, (8), leads to contradiction.

2. One, however, might reply by replacing (8) with (9):

(9)   We shall diagnose a change of meaning if each term employed by T has extension zero when employed by $\bar{T}$.

(9) presupposes only that there are terms *typographically* common to T and $\bar{T}$. This presupposition is not pernicious. It is a commonplace that one of the things that changes least in scientic revolutions is basic vocabulary. (9) does not presuppose that the *concepts* of T are carried over, and common to, $\bar{T}$. It would therefore seem to avoid my objection to (8). Assuming 'S_1' ... 'S_k' ... are all the terms used by T and assuming that each also occurs in theory $\bar{T}$, (9) thus makes the claim that if 'S_1' ... 'S_k' ... have extension zero in $\bar{T}$ then their meaning has undergone a change since as used by T these same typographic terms presumably do not have extension zero. Interestingly, (9) would seem to be well in keeping, not with a contextualist theory of meaning which stresses the importance of sense or connotation, but rather, with a more conservative empiricism which stresses reference or extension (e.g. [23], Chapter 1; [64], pp. 47–48, 132; [73], pp. 57, 62).[12]

This emendation of Feyerabend's criterion, although an improvement over (8), does not, however, support his further conclusion that there has been a change of meaning in the transition from classical mechanics to relativistic mechanics. This is because the predicate 'has mass' employed by classical mechanics (T) does not have extension zero in its employment by general relativity ($\bar{T}$), for in general relativity some things do have mass. Further, since Feyerabend speaks only of classes of "objects and events" (9) would direct attention only to the extensions of predicate terms to the neglect of both operation and magnitude terms. The extension of magnitude terms, for example, is a class of *ordered pairs* with objects as the first, and *numbers* as the second, entry of each element. That is, (9) would sanction a wholesale diagnosis of change of meaning for operation, magnitude, and predicate terms if change were to occur in the meanings of only the latter sort of terms.

CONCLUSION

I discussed at the outset of this Chapter two central theses of meaning variance theorists (I). I argued there that, contrary to their suggestions, one can maintain the first without maintaining the second. I then examined and found to be unsuccessful three types of arguments and criteria (Sections II, III, and IV) which purport to establish different kinds of meaning variance in scientific transitions. I conclude that the proposed arguments and criteria for radical meaning variance have failed.

NOTES

[1] I will present further arguments for this methodological point in Chapter 3. I will there discuss, among other things, proposed ways for evaluating the merits of different theories independently of experience. And in Chapter 4 I will present additional arguments to the effect that observational invariance should occur and has in fact usually occurred during scientific transitions.

[2] [12], p. 218; [16], p. 180; [27], pp. 57, 154; [28], p. 66; [42], pp. 96–102; [71], p. 90; [82], pp. 13–16, 168–170; [88], paragraph 43, p. 20.

[3] [14], pp. 29, 59; [15], p. 36; [27], pp. 7, 54–58, 61, 96–99, 154–156; [28], p. 38; [34], p. 102; [35], pp. 51–52; [42], pp. 145, 148–149; [81], p. 162; [82], pp. 18, 20–21; [83], p. 57.

[4] Cf.: [15], p. 30; [16], pp. 179–180; [27], pp. 54–58, 61, 154–156; [35], pp. 47, 51–52; [42], pp. 96–102, 147–149; [71], p. 90; [81], p. 162; [82], pp. 20–21, 35–38; [83], pp. 55–59. Even Shapere seems to feel that thesis (1) entails thesis (2); cf. [78], pp. 54–56, 67.

[5] For a good discussion of some of these difficulties cf. Goodman, [24], pp. 67–71.

[6] Mates admits that we need "a different and more general criterion for the synonymity of expressions occurring in different languages" ([55], pp. 119–120).

[7] Quine, unlike Goodman, expresses doubts about whether there exists a criterion of synonymy at all ([64], pp. 48, 63). To this extent the approach of radical meaning variance theorists would be incompatible with Quine's position, for they employ such criteria in their arguments for radical meaning variance (cf. II, III, IV).

[8] Cf.: Feyerabend, [14], pp. 55–57, 80–81; [15], pp. 14–15; [16], pp. 168–170; Hanson, [27], pp. 150–157; [28], pp. 69–70; Hesse, [35], p. 51; and Kuhn, [42], p. 101.

[9] That there can be, and usually is, a common domain is what we argued in Chapter 1. We will return to this in Chapter 4 where we shall adduce additional support for observational invariance.

[10] We shall briefly return to this topic in Chapter 3, Section II and VI.

[11] For a discussion of the variety of purposes to which a scientific theory can be employed cf. [40], [56].

[12] It is unfortunate that Scheffler's, otherwise admirable, recent book ([73], especially Chapter 3), which explicitly attacks the radical meaning variance position, stresses the importance of extension for science presumably as an answer to radical meaning variance theorists and yet fails to discuss anything like either (8) or (9).

CHAPTER 3

THE METHODOLOGICAL UNDESIRABILITY OF ADOPTING A POSITION OF RADICAL MEANING VARIANCE

In Chapter Two I argued that the arguments for radical meaning variance have failed. In the present Chapter I wish to argue that the radical meaning variance position in itself is unreasonable. That is, I wish to argue that it is unreasonable to maintain that transitions from one scientific tradition to another force a change in the meanings of the terms employed which is radical enough to preclude the possibility of comparisons of scientific theories from different traditions through appeal to, and on the basis of, some sort of shared meaning of the terms employed. The position has several methodologically unacceptable consequences which are not, as has been argued, avoidable. As an aid in noting some of these I feel that it would first be profitable to examine some examples which have been put forward to illustrate and suggest the position.

I

I wish to suggest here that the radical meaning variance thesis is not only unsupported (Chapter 2) but false. Several examples have been put forward to illustrate and confirm the radical meaning variance thesis. I wish to argue, instead, that these very examples suggest the falsity of this thesis.

According to Hanson; Brahe and Kepler do not mean the same thing by 'sun' ([27], pp. 7, 23; cf. also pp. 96–97). Immobility is built into Kepler's concept of the sun; 'is circled by the earth', is part of his definition of 'sun'. Similarly, mobility is built into Brahe's concept of the sun; 'revolves around the earth' is part of his definition of 'sun'. Now, if this is true, then

(1)   The sun is immobile.

must be, for Kepler, true in virtue of the meanings of the terms involved (cf. [27], pp. 96–97). Similarly, (2) must be true in the same way for Brahe:

(2)   The sun is mobile.

(2) is, for Kepler, self-contradictory since he holds (1) to be *a priori*. And (1) is, for Brahe, self-contradictory since he holds (2) to be *a priori*.

Such a situation is, however, implausible, for the following two reasons.

The first is that the *issue* of helio-centrism versus geo-centrism is one which was resolved by a *contingent* and *a posteriori* truth. The above, however, would preclude this.

The second is that both Kepler and Brahe performed empirical experiments to confirm and substantiate their theories (helio-centrism and geo-centrism respectively). This itself supports the claim that, for Kepler, (1) was contingent, and *not a priori*; one does not perform empirical experiments to test an *a priori* truth just as one does not perform empirical experiments to test the truths of pure mathematics. Similarly, for Brahe, (2) was contingent and not *a priori*. Part of the justification we have for saying a particular truth is *a priori* for someone is that he does *not* attempt to test or confirm this truth through experiment. And in this case, such justification is lacking because Kepler and Brahe each did attempt to test or confirm (1) and (2) respectively through experiment.[1] Such attempts at empirical confirmation, moreover, do provide us with some justification for the counterclaims that both Kepler and Brahe *did* regard (1) and (2) as contingent and empirical truths, that Brahe had *not* built mobility into his concept of the sun, and that Kepler had not built immobility into his concept of the sun.

Similar and equally implausible examples abound. According to Feyerabend ([14], pp. 77–80; cf. also Toulmin, [82], p. 130) current scientists in their use of the word 'temperature' mean something radically different from what Galileo meant. Their scientific differences are held to hinge on questions of meaning rather than on questions of fact. The evidence Feyerabend gives is that we have abandoned the proposition "the temperature shown by a thermometer is not dependent upon the chemical constitution of the fluid used" which he takes to be constitutive of the Galilean concept. Feyerabend would hold that this was built into Galileo's concept of temperature. Again, according to Feyerabend ([14], pp. 80–81; [15], pp. 14–15; [16], pp. 168–170) and Kuhn ([42], p. 101) the meaning of 'mass' (among other terms) has radically and incommensurably changed meaning in the transition from classical to relativistic mechanics. For Kuhn and Feyerabend dependence on velocity and convertibility with energy are built into the relativistic concept of mass.

Smart seems to consider such an identification of the meaning of a term with the particular theory containing the term to be the most interesting part of Feyerabend's position ([81], pp. 162–163). Smart is quite sympathetic to such a view; he offers us a supporting example that 'chimney pot' has changed its meaning as a result of modern physics.

The philosophical interpretations of each of these examples are implausible. The reasons are similar to those which we noted above in discussing Hanson's example. None of these examples illustrates or confirms the truth of the radical meaning variance thesis. Rather, I would urge, they suggest the falsity of this thesis. In Chapter Two I examined, and found to be unsatisfactory, the reasons, arguments, and criteria used to support the radical meaning variance thesis. If I am correct in this section then we have, in addition, at least some ground for considering this thesis itself to be implausible.

There are, however, deeper objections to the radical meaning variance thesis. To these we now turn.

II

A. The first methodologically undesirable consequence of the doctrine of radical meaning variance is that if it were true then no theory could contradict or be consistent with another.

On the radical meaning variance view when a new theory T' emerges to replace an old one T, the terms involved – both theoretical and observational – will change in such a way that there will be an "elimination of the old meanings" ([14], p. 29). The same term, although employed in both cases, will express two radically different, "incommensurable" concepts. In particular, if this is true then T' and T could not contradict each other or be mutually inconsistent. Assume that T' and T did contradict each other or were mutually inconsistent. There will then exist two statements **A** and **B** which are expressible in, and hold relative to, T' and T respectively. **A** and **B**, further, will contradict, or be inconsistent with, one another. Ordinarily this would mean that in T', **A** entails $\sim$**B**. With respect to our two statements **A** and **B** this is to say that what is denied by the one must be entailed by what the other asserts. Thus, a statement (either $\sim$**B** or **B** will do here) expressible in each theory will have the same meaning in each theory; this in turn is to say that the theories must have some common meaning, in contradiction to the radical meaning variance doctrine as expressed by Feyerabend, Hanson, Hesse, Kuhn, and Toulmin.[2] Alternately put, if the radical meaning variance doctrine is correct and if in T', A entails $\sim B$ then A and B cannot be statements expressible in T' and T respectively.

Following Achinstein ([1], p. 499), let us consider the Bohr theory of the atom. It claims that electrons revolve about the nucleus so that their angular momentum is quantized. It further claims that energy is quantized and is radiated or absorbed by the atom only when an electron jumps from one stable orbit to another.[3] When the Bohr theory claims that angular momentum and radiant energy of electrons must be quantized and cannot have continuous values, it thereby denies the claim of classical electrodynamics that angular momentum and radiant energy of electrons can have continuous values. In Bohr's words:

> ... *the essential point* in Planck's theory of radiation is that the energy radiation from an atomic system does *not* take place in the continuous way assumed in the ordinary electrodynamics, but that it *on the contrary* takes place in distinctly separated emissions. ([7], p. 4, all my italics)

It would be hard to preserve the possibility of such denials given radical meaning variance. The terms in the Bohr theory would be held to have radically different meanings from those in classical electrodynamics. Indeed, they are held to express "incommensurable" concepts. Pending the emergence of a new and viable sense of 'denial' or 'incompatible' different from the notions of contradiction or inconsistency such denials would be mere illusions.

Feyerabend does attempt to provide a new sense for 'denial' and 'incompatible' which does not appeal to, or presuppose, shared meaning. We will soon examine this. In any case, Bohr's assertion that energy radiation from an atomic system is not continuous could not be *contrary* to the classical claim; both Bohr and a classical theorist would be held to use words and theories with radically different and "incommensurable" meanings; thus they could not contradict one another. Bohr's above remarks would therefore be false; he asserts that the *essential point* of his and Planck's view is *contrary* to the classical assumption.

Similar examples abound. Lavoisier denied, and Priestley asserted, that "respirable" air was dephlogisticated. Lavoisier denied, and Priestley asserted, that metals were a combination of calx and phlogiston.[4] Lavoisier himself writes as follows:

... Priestley ... has sought to prove, by very ingenious, very delicate, and very novel experiments, that the respiration of animals has the property of phlogisticating the air, like the calcination of metals and several other chemical processes, and that air only ceases to be respirable at the moment when it is overcharged and in some way saturated with phlogiston.

However probable the theory of the celebrated physicist may have appeared at first sight, however numerous and well-performed the experiments on which he sought to establish his theory, I must own that I have found it to be *contradicted* (my italics) by so many phenomena that I have thought myself justified in calling it in question; I have consequently worked on another plan and I have been irresistibly led by my experiments to conclusions quite other than his... Priestley's experiments... all prove to be in favour of the opinion I am going to expound in this memoir. ([44], p. 88)

Again,

...pure air, respirable air... which forms only a quarter of atmospheric air... Priestley has very *wrongly* (my italics) called *dephlogisticated air*. ([45], p. 92)

Again, pending the emergence of a new and viable sense of 'denial' or 'incompatible', different from the notions of contradiction or inconsistency it would be difficult to preserve the possibility of such denials on the radical meaning variance account. The terms used by Lavoisier would be held to have radically different, "incommensurable" meanings from those used by Priestley. In any case, Lavoisier's assertions could not *contradict* those of Priestley. Hence, Lavoisier's claim that his own assertions did *contradict* those of Priestley would have to be adjudged false.

Such consequences would render many assertions false presumably because the scientists who made them were not familiar enough with the meanings of the terms they employed in their own theories to recognize that these meanings were radically different from the meanings of the terms employed by their competitors. Yet, it is usually the principal advocates and originators of a theory who, above all, are taken to understand the meanings of the terms they employ and of the terms their competitors employ. Indeed, it is they who use these terms daily and it is they who engage in the prolonged "paradigm debates".

The consequence – that no theory could either contradict or be consistent with another – is further directly opposed to one of the principal methodological reasons that motivates Feyerabend to hold the radical meaning variance position in the first place, namely the "principle of proliferation: Invent, and elaborate theories which are *inconsistent* (my italics) with the accepted point of view, even if the latter should happen to be highly confirmed and generally accepted" ([18], pp. 223–224). This principle is central to Feyerabend's positive methodology and is a "main consequence" of his "abstract model for the acquisition of knowledge" ([18], p. 223). This principle, under the name of 'theoretical pluralism', "is assumed to be an *essential feature* (his italics) of all knowledge" ([15], p. 6). But it is precisely his radical meaning variance position which precludes two different theories from being *inconsistent*. Rather than being "an *essential feature* of all knowledge", the principle of proliferation would therefore be an *impossible feature* of all knowledge. And on a normative interpretation it would, in effect, call for the impossible given the doctrine of radical meaning variance. Prescriptions for the impossible, though not strictly contradictory, are somewhat awkward and become especially suspect when employed within a *scientific* methodology. Given all this I therefore find it quite ironic that Feyerabend puts forward precisely this methodological principle as a *reason* for his espousal of radical meaning variance (cf., e.g. [15], pp. 7–8, 30).

The fact that different theories could not contradict one another if the radical meaning variance thesis were true also destroys another of the reasons or arguments usually adduced to establish such variance in the first place. Feyerabend and Kuhn conclude that terms occurring in T (say, classical mechanics) occur with radically different meaning in T' (say, relativistic mechanics) from the fact that the basic principles of T and T' contradict one another, or from the fact that T and T' lead to inconsistent consequences in certain domains ([14], p. 82; [15], pp. 13–16; [42], pp. 100–101). However, given the above consequences we have noted, this sort of argument is self-destroying; the truth of the conclusion is incompatible with the truth of the premises. If radical meaning variance indeed occurred in a particular scientific transition then no evidence for it from this sort of argument could exist. That is, T and T', being theoretically and observationally incommensurable, could have neither mutually *contradictory* basic principles nor mutually *inconsistent* consequences in any domain.

B. Feyerabend has, partially in effect, recently ([18], pp. 231–234) replied to such criticism.[5] He claims that disagreement between two incommensurable theories can be established without appealing to the meanings of terms, and hence without assuming either overlap in meaning or any "minimum of similarity in meanings" ([18], p. 233). Thus, he feels that "it *is* possible to use incommensurable theories for the purpose of mutual criticism" ([18], p. 234) since "there may be empirical evidence against one and for another theory without any need for similarity of meanings" ([18], p. 233). He feels that this could occur in at least two ways.

The first is as follows ([18], pp. 232–233):

(3)   Two theories are incompatible (i.e. disagree) if they are not isomorphic.

Given the correctness of (3), Feyerabend could presumably then use the two non-isomorphic theories for the purpose of mutual criticism by testing each, adopting the most highly confirmed, and discarding the other. Feyerabend may have in mind either isomorphism between the systems of elements described by the respective theories or between the systems of terms and sentences (both theoretical and observational) used by the respective theories. And he correctly notes that lack of isomorphism can be established "without bringing in meanings at all" ([18], p. 232).

(3), however, is false. Lack of isomorphism of either kind is not a sufficient condition for either incompatibility or disagreement between theories.[6] The systems of elements described by sociological theory and quantum theory respectively, are not isomorphic. Nor are the systems of terms and sentences used by these respective theories isomorphic. Nevertheless, sociological theory and quantum theory are not incompatible or in disagreement.[7] (3) is, therefore, false. It is unsatisfactory and does not enable two incommensurable theories which are radically different in meaning to be incompatible with, contradict, or disagree with, one another. The objections that I raised in Section IIA thus remain.

Feyerabend, however, presents a second reply to these sorts of objections. He feels that there is a second way in which two incommensurable theories may disagree and be judged against one another. His reply is as follows:

...Secondly, both theories may be able to reproduce the "local grammar"... of sentences which are directly connected with observational procedures. In this case the utterance of one of the sentences in question in accordance with the rules of the local grammar, or the utterance of a local statement, as we shall say, can be connected with *two* "theoretical" statements, one of T, and one T' (*sic.*) respectively (theoretical statements, corresponding to a given local statement S will be called statements associated with S within the theory in question). We may now say that the empirical content of $T' >$ the empirical content of T, if for every associated statement of T there is an associated statement provided by T', but not vice versa. And we may also say that T' has been confirmed by the very same evidence that refutes T if there is a local statement S whose associated statement in T' confirms T' while its associated statement in T refutes T. Similar remarks apply to the question of independent evidence. ([18], p. 233)

Although vaguely formulated, I am not unsympathetic to the tenor of Feyerabend's above procedure for choosing between two different theories. As a *reply*, however, to our objection (IIA) to the doctrine of radical meaning variance it is unsatisfactory; indeed it is incompatible with that doctrine. Assume that the utterance of "local" statements "which are directly connected with observational procedures... can be connected with *two* 'theoretical' statements, one of T, and one [of] T' respectively" in such a way as to both determine the "empirical content" of, and serve to confirm or falsify, T and T'. If this is so then indeed these local statements would seem to be nothing other than what has traditionally been called "neutral observation statements"; but, as held by Feyerabend, Hanson, and Kuhn, the radical meaning variance doctrine is understood to include radical change in the meaning of observational statements and terms, as well as the theoretical ones. It seems, therefore, that Feyerabend has introduced something very similar to neutral observational statements, whose meaning he assumes to be invariant with respect to the two theories which are to be compared, in order to preserve the doctrine of radical meaning variance; yet this very doctrine is such that it denies the possibility of such statements (cf., e.g. some of Feyerabend's other remarks: [14], pp. 28–29; [15], pp. 15–17; [16], pp. 213–214).

Such "local statements" would further, on his above reply, seem to formulate the common facts neutral to theoretical controversies and against which we can compare and test theories. This would entail that there is some common overlap of experience between competing theories, namely the facts or the "observational procedures" (e.g. "looking", cf. [18], p. 259, n. 32). But Feyerabend's deeply held view of observation and experience – "each theory will possess its own experience, and there will be no overlap between these experiences" ([16], p. 214) – would deny just this.

Such neutral experience, on the above reply, insofar as it is used to compare the two theories, would also seem to entail that there exists some sort of stability of reference or extension for some of the terms and sentences employed by each theory; in this sense, then, some "meaning" will *not* have changed for the purpose of comparing different theories, in direct opposition to the doctrine of radical meaning variance.

If we have been correct in IIA and IIB then no successful way has been provided by which different theories could either agree or disagree with one another if the doctrine of radical meaning variance is correct.

III

The second methodologically undesirable and unacceptable consequence of the doctrine of radical meaning variance is an extension of the first. It can be stated in the following way. If the doctrine were true then each scientist would be effectively isolated within his own system of meanings. Each of these meanings would be radically different from those of scientists within other traditions or from those of scientists holding other theories. True communication in any sense – not just agreement or disagreement – from one such system to another would be impossible. There could be no intelligible converse – not merely a lack of intelligible mutual in-compatibility or inconsistency – between scientists holding different theories. Each, as Scheffler notes ([73], p. 46), would be "trapped in the web of his own meanings."

Kuhn maintains that *competing* ([42], p. 149) paradigms are addressed to radically different problems. They incorporate radically different standards and even radically different definitions of science. They are based on radically different meanings. They operate in radically different observational worlds.

But if this is so then in what sense could such paradigms be said to be in competition? How could they be either *rivals* or *alternatives*? To main-tain that they are in competition is to place them within *some* common framework which has comparative and evaluative standards applicable to both. It is to consider them as oriented in somewhat different ways toward the same purposes and scientific goals. It is to consider them as dealing in somewhat different ways with at least some common referential situations.

And it is the invention of *alternatives* for the purposes of mutual criticism which is central to Feyerabend's own positive methodology. Yet, given his espousal of radical meaning variance, it is just this which would be impossible. Feyerabend claims:

You can be a good empiricist only if you are prepared to work with many *alternative* (my italics) theories rather than with a single point of view and 'experience.' ([15], p. 6)

Given the radical meaning variance position, two different theories are radically different in meaning. It is thus hard to see how they could function as *alternatives* to each other or serve to *criticize* ([15], p. 6) each other, just as sociological theory and quantum theory which are radically different in meaning are neither alternatives to each other nor serve to criticize one another. Feyerabend puts forward precisely the methodological goal of mutual criticism using alternative theories as a *reason* for his espousal of the radical meaning variance position (cf., e.g. [15], pp. 7–8, 30). He thinks that adopting the radical meaning variance position enables us to come closer to and more fully reach his goal. My entire argument in this section and in the preceding (II) implies that precisely the reverse is the case.

On the radical meaning variance view, theory displacement in science is held to affect observational as well as theoretical categories and notions; it follows that apparent sharing of observational terms by theoretical opponents is really a delusion. There are perhaps common symbols and sounds, but no non-trivial common meanings. If we are to understand another's observational or experimental claims it is held that we must first accept his theory. Further, it is held that if we are to understand another scientist's language we must share his theory; this means we must share certain deep features of his thought-world, his outlook, expectations, and beliefs, for it is only in terms of these, it is claimed, that his language can be rendered intelligible.[8] True communication between holders of different scientific theories thus becomes impossible. For a scientist of one tradition to significantly converse with an "opposing" theorist from another tradition on neutral ground becomes impossible. Conversation in short *requires* not open-mindedness, but closed-mindedness, and *presupposes* conversion (cf., e.g. Kuhn, [42], p. 149). Radical meaning variance is not, as Feyerabend thinks (cf. [15], pp. 18–19, 25–33, 37–38), methodologically desirable. It leads to isolation. And this, from a methodological viewpoint, is surely undesirable.

IV

There is a third methodologically undesirable and unacceptable consequence of the doctrine of radical meaning variance. If the doctrine were true it would be difficult to see how one could learn a new theory.[9] He could not learn it by having it explained to him using any scientific terms whose meanings he understood before he learned the new theory. Consider the terms 'mass', 'velocity', and 'energy' which are used in relativistic mechanics. The meaning of each of these terms, given the doctrine of radical meaning variance, is theory-laden. As Hanson would say, "The entire conceptual pattern of the game is implicit in each term" ([27], p. 61). If so, then, in order to know what the terms of relativistic mechanics mean, I must know relativistic mechanics or at least its central principles. One of these central principles can be roughly expressed like this:

(4)   Mass is a function of velocity and is convertible with energy.

But it is hard to see how I could know what (4) asserts unless I already, to some degree at least, know what the terms 'mass', 'velocity', and 'energy' mean. Radical meaning variance theorists would hold that to learn what the terms mean I must learn the theory. To learn the theory, however, I must learn its central principles. But to learn the latter, it would seem that I must know what the terms involved mean. Given radical meaning variance, this circularity is vicious. In trying to circumvent it, it is useless to appeal to what the terms mean in other, different theories; what they mean in any two theories is held to be radically different and incommensurable. Nor could we explain what they mean by using terms whose meanings are independent of any theory; on the radical meaning variance doctrine there are supposed to be no theory-neutral terms.[10] Therefore, I could not use any terms whose meanings I previously understood in order to learn the meanings of the terms in relativistic mechanics. Most scientific textbooks would therefore end up useless in principle if the radical meaning variance account were correct. Yet in actual practice − by means of scientific textbooks and verbal scientific lectures – most terms in a new theory are successfully explained to those learning the theory by using terms whose meanings the learners already know.

The situation, however, is still worse. Could I learn the meanings of the terms in relativistic mechanics extralinguistically? If I could do this then I would have to "watch what those who use the theory do in their laboratories, the sorts of items to which they apply their terms", etc. ([2], p. 97). As Achinstein correctly notes ([2], p. 97), I would then have to "learn each new theory like a child first learning language (rather than like someone learning more of his own language, or a second language after learning a first one)". To successfully learn relativistic mechanics I would have to observe the non-verbal behavior of those who already know it along with the phenomena such people may point out. However, if the radical meaning variance view of observation (cf. Chapter 1, Section I) were correct, I could indeed *not* learn relativistic mechanics even in the latter, extralinguistic way. They would hold, contrary to logical empiricists or operationalists, that the observing of such phenomena already presupposes (cf. Chapter 1, Section VD) that one understand and accept relativistic mechanics; and it is just this which the student-learner is not yet equipped to do.[11]

Thus, given the general radical meaning variance position, it is difficult to see how one could, in any sense, learn a new theory.

V

There is a fourth methodologically undesirable and unacceptable consequence of the doctrine of radical meaning variance. If the doctrine were true, no scientific theory could be tested or falsified by any observations or observation reports.

A. The claim that the meaning of both observational and theoretical terms occurring in a scientific theory should depend *entirely* on the central principles of that theory suggests, as Achinstein notes ([2], p. 96), that these principles in effect say what the terms mean. "If so, then such principles would always be" defensible by appeal solely "to the meanings of their constituent terms" ([2], p. 96). The claim (4) of relativistic mechanics could then be defended by appeal solely to the fact that 'mass', 'velocity', and 'energy' mean just those sorts of things which make (4) true. However, it is not only the central theoretical principles of scientific theories which end up true in virtue of the meaning of their constituent terms, given the radical meaning variance position. Most, if not all, of the other, less central theoretical principles do as well, for these are held to follow from the central ones. And the more observational and less theoretical statements of scientific theories would also seem to end up defensible in this way, given the radical meaning variance position (cf. Section I of this Chapter). This is because observational terms are held to have no special status; they are not held to have independent observational or experimental meaning; they are held to be defined entirely by means of the theory and its central principles. It would, as Achinstein notes ([2], p. 96), indeed be hard to see how any observed item could be *described* as failing to satisfy the postulates of a scientific theory given that any description must use terms and statements whose meaning flows entirely from those postulates. We would have to describe any bachelor as 'unmarried' if we indeed do describe him as a 'bachelor'. Just so, given the radical meaning variance account, we would have to describe light near the sun as 'travelling in straight lines' if we indeed to describe it as 'light'.[12] Each of a scientific theory's claims – both observational and theoretical – would end up true in virtue of the meanings of the terms employed. It would therefore follow that no scientific theory would be falsifiable by means of observations or observation reports, just as the statement "All bachelors are unmarried" is not falsifiable by means of observations or observation reports.

The radical meaning variance doctrine as usually expounded would thus seem to deny that there is a "truths of language – truths of fact" distinction. And this denial would not merely be in the sense that there is no sharp distinction. For radical meaning variance theorists this denial would seem to have to take the form of a claim that nothing is synthetically true, that all truth or falsity is linguistic (cf. Chapter 1, Section III). Radical meaning variance theorists might reply to this objection by maintaining that although the central principles of a scientific theory are true in virtue of the meanings of their constituent terms, the observation and existence claims are not. They might hold, as Achinstein points out ([2], p. 96), that "the claim that there exist atoms satisfying the principles of the Bohr theory will be empirical". On this reply the postulates of a scientific theory would remain true in virtue of the terms involved; but "the question whether anything satisfies these postulates" becomes "empirical". Although all observation claims and reports will presuppose the theory and its central principles, it will thus be possible for observation claims and reports to be false; they will not all be true solely in virtue of the meanings of their constituent terms. Scientific theories thus become empirically testable by reference to observation; false observation claims and reports could then provide a basis for the rational rejection of a scientific theory and for the rational acceptance of a new theory which is inconsistent with the old.

Such a reply, however, will not work. In the first place, one might historically object to both of these versions of reconstructing scientific theories. They make many statements that one would normally take to be empirical – statements scientists in fact treat as such – now defensible "simply by appeal to the meanings of their terms" ([2], p. 96). This is implausible for the objections I raised at the outset of this Chapter (cf. I) in connection with Hanson's 'sun', Feyerabend's 'temperature', and Kuhn's and Feyerabend's 'mass'. I would agree with Quine to the extent that I think that none of the central principles proper to a scientific theory are either analytic or true simply in virtue of the meanings of the terms involved. I would also maintain that none of the observation claims are analytic; they are empirical and can be false.

There are, however, two more serious problems with respect to the above reply and the radical meaning variance doctrine generally. First, if either is correct, then observation claims and reports could not lead to the rational rejection of a scientific theory; that is, they could not falsify a theory. Nor, secondly, could they lead to the rational acceptance of a new theory which is mutually inconsistent with the old.

Consider the first point. According to Hanson, Feyerabend, Kuhn, and Toulmin, all terms employed by a scientific theory are laden with that theory; this includes those terms employed in describing the observations made for the purpose of its confirmation and test. They claim that any observation and its description O presupposes, and is laden with, the theory T (cf. Chapter 1, VD). On Van Fraassen's account of the concept of presupposition ([86], esp. pp. 137–138), the only developed account of this notion known to me, it then follows that O is neither true nor false unless T is true. And from this it follows that if O is true then T is true and if O is false then T is true. Thus, we cannot maintain that O is true and T is false; nor can we maintain that O is false and T is false. The latter result is rather remarkable; it implies that false predictions could not be used as a basis for denying T; if observation report O fails to satisfy T's observation claim or prediction P, we could not use this as grounds for claiming that T is false. The situation, however, is still worse. No observation report O, whether true or false, arrived at employing T could be used to reject T; if O presupposed T then whenever O would be true, T would be true, and whenever O would be false, T would be true. The only way out of this dilemma is to maintain that reports of observations are neither true nor false. But this is tantamount to denying that science is an empirical and cognitive enterprise. In any case, we would have to regard as neither true nor false *all* reports of observations made during former times, if we regarded as false the theory employed during those times. In particular, the observation reports of Brahe, Kepler, Newton,

Leverrier, Priestley, Gibbs, and Michelson-Morley would have to be regarded as neither true nor false.

But perhaps Feyerabend and Kuhn ([42], p. 77) would agree that without the help of other theories, no single theory could be falsified. As Feyerabend puts it:

One most important point of agreement [between Kuhn and Feyerabend] is the emphasis which both of us put upon the need, in the process of the refutation of a theory, for at least another theory. ([14], p. 32)

... there ... exist facts that cannot be unearthed except with the help of alternatives to the theory to be tested and that become unavailable as soon as such alternatives are excluded. ([16], p. 175)

Both the relevance and the refuting character of many decisive facts can be established only with the help of other theories that, although factually adequate, are not in agreement with the view to be tested ... Empiricism demands that the empirical content of whatever knowledge we possess be increased as much as possible. ([16], p. 176)

This, however, is also implausible if the doctrine of radical meaning variance is correct; indeed it is inconsistent with that doctrine. A different theory T' could not be used to show that observations exist which do not satisfy T. The concepts of T' would be held to be incommensurable with those of T. Thus, satisfied prediction statements expressible in T' could not be used, as Feyerabend wishes them to be used, to show that items exist which satisfy the negations of prediction statements expressible in T. If they could then T and T' would not be incommensurable (cf. Sections II and III), contrary to the radical meaning variance position.

The radical meaning variance position would make a scientific theory T irrefutable and untestable by any observation that could be described either by T or by another theory T'. Nor could observations made independently of any theory refute a scientific theory, since for Feyerabend, Hanson, Kuhn, and Toulmin, observations must always presuppose some theory; they are theory-dependent. If O is made presupposing T, then T could not be refuted; nor could T be refuted using T', since T and T', being incommensurable, could not disagree (cf. II). And disagreement over empirical content is precisely what Feyerabend would desire for his incommensurable alternatives (cf. above, [16], p. 176). In no sense then, given radical meaning variance, could scientific theories be called 'empirical'. In no sense could they have what Feyerabend wants them to have, namely "empirical content" (cf. e.g. above, [16], p. 176).

There is another problem. No true or false observation report O which presupposes a theory T could serve as a basis for the rational acceptance of a new theory T' which is mutually inconsistent with T. If we accept T', we would have to deny T, given that T and T' are mutually inconsistent. If we deny T, we would have to maintain that O is neither true nor false. If we maintain that O is neither true nor false it is hard to see how O could serve as a basis for the rational acceptance of T'. In particular, if we accept relativistic physics (T') we would have to deny classical physics (T) since these two theories are mutually inconsistent. If we deny classical physics then we would have to maintain that the observation reports (O) of the Michelson-Morley experiment, of Leverrier's observations of Mercury's rotation round the sun, etc., were neither true nor false. Being neither true nor false it is hard to see how they could have served as a basis for the rational acceptance of relativistic physics over classical physics. These observation reports, however, *are* generally considered by scientists to be correct and true, and are part of what led to the rational acceptance of relativistic physics over classical physics (cf., e.g. [54], pp. 542–547).

B. The fact that no theory could be tested or falsified by observations, given the radical meaning variance position, is, further, directly opposed to Feyerabend's own "abstract model for the acquisition of knowledge... which... has as its *aim* (his italics) *maximum* (my italics) testability of our knowledge" ([18], p. 223). The aim of this model is indeed a principal reason which prompts him to hold the radical meaning variance position in the first place. He maintains that without recourse to alternative theories which are radically and incommensurably different in meaning from T we could not "unearth" many facts relevant to the testing of T (cf., e.g.: [15], pp. 3–8; [16], pp. 174–177).[13] If, however, our above conclusion is correct then his aim of maximum testability is incompatible with precisely that position he puts forward to attain it. His aim would be unrealizable given his radical meaning variance position. *Any* observation would be precluded from falsifying or testing a scientific theory. Thus, we would have, if anything, *minimum*, not "*maximum* testability of our knowledge." For this reason it is also hard to see how Feyerabend can regard "Popper's admirable *Logic of Scientific Discovery*" as "both the starting point and motive force" of his investigations ([14], pp. 31–32). Indeed, Popper expresses one of his most central claims in that work when he asserts that he:

shall certainly admit a system as empirical or scientific only if it is capable of being *tested* by experience ... the falsifiability of a system is to be taken as a criterion of demarcation ... *it must be possible for an empirical scientific system to be refuted by experience.* ([61], pp. 40–41, his italics)

Feyerabend could not consistently regard such a view as the "motive force" for his espousal of the radical meaning variance position; this position is incompatible with a view such as Popper's.

C. Our conclusion in *A* is that given the doctrine of radical meaning variance no scientific theory could be tested or falsified by any observation or observation report. This conclusion, however, might be thought to be neither undesirable nor unacceptable. In particular, Kuhn, unlike Feyerabend and Popper, might look on our conclusion in this way. Kuhn maintains that scientists:

do not ... treat anomalies as counterinstances, though in the vocabulary of philosophy of science that is what they are ... No process yet disclosed by the historical study of scientific development at all resembles the methodological stereotype of falsification by direct comparison with nature. ([42], p. 77)

Kuhn's reasons for this claim appeal to the behavior of scientists. Exactly what his reasons are, is quite difficult to determine, and even more difficult to concisely express. If, however, Purtill's interpretation ([62]; cf. also Scheffler's interpretation which is similar, [73], pp. 18–19, 86) of Kuhn is correct then three of Kuhn's reasons can be stated as follows:

(5)   Many scientists do not change paradigms after the initial successes of a new paradigm.

(6)   Some scientists never do change paradigms, even after most of their colleagues have done so.

(7)   Many of the early supporters of a new paradigm have no more than a hope that it will explain previously unexplained phenomena.

(5) and (6) hinge on the apparent resistance of paradigms to falsification. Interpreted thus, Kuhn would feel that if paradigms were really subject to test then they should be expected to fall with the discovery of counterinstances. Each reason would be held to result in the conclusion that the criterion of falsifiability is a myth. Paradigm changes would not be justifiable, since the behavior of scientists would show that mere persuasion rather than proof is going on. On Purtill's interpretation Kuhn would say that in the past claims like (5)–(7):

have most often been taken to indicate that scientists being only human, cannot always admit their errors even when confronted with a strict proof. I would argue, rather, that in these matters [the "transfer of allegiance from paradigm to paradigm"] neither proof nor error is at issue. ([42], p. 150)

Following Purtill, let us examine these reasons, (5)–(7), for Kuhn's conclusions that the criterion of falsifiability is a myth and that paradigm changes are not justifiable. These reasons tend to look impressive when supported by Kuhn's wealth of historical examples. In my opinion, however, they do not support his conclusions. Let us turn to (5) and (7). Although they conflict, Kuhn feels each supports his conclusion. Consider the Copernican Revolution. Kuhn would conclude that this paradigm change is not rationally justifiable and that the criterion of falsifiability here is a myth. One reason he would give is (7). At first, Copernicus' theory presumably had no greater explanatory or predictive success than its rivals and thus its early supporters had no more than a hope that it was the best available theory. However, another reason Kuhn would give is (5). Scientists at first hesitated to adopt Copernicus' theory. That there may be a connection between these points seems to escape Kuhn's attention, as Purtill notes ([62], p. 57). One could say that as the superiority in prediction and explanation of Copernicus' theory became increasingly apparent it became increasingly accepted.

There is more to be said about (7). Many scientists do back a theory early before there is definite explanatory or predictive success to support it over its rivals. But what *need* be wrong with this? These scientists were hoping that they were backing a winner and some of these hopes were fulfilled. The notion of the *acceptance* of a scientific theory or hypothesis is critically ambiguous. It may mean holding them to be true. However, it may also mean regarding them seriously enough for their heuristic employment as leading principles of research and experimentation, without a definite commitment as to their truth value. The acceptance, in either sense of 'acceptance' of a relatively unsupported theory or hypothesis is compatible with the existence and acknowledgment of criteria of controlling tests, such as the criterion of falsifiability, to which future experience will subject the theory. Many scientists, however, have backed losers, and later found that their early hopes of success were not to be fulfilled. Yet this does not imply that their early hopes were not justified at the time. If it did, there would be a dangerous tendency to write history from the point of view of hindsight. For example, it is *now* thought that Copernicus' theory was the best available at the time; thus, somehow its early support was justified (which incidentally contradicts the use to which Kuhn puts (7)), and its initial opposition was not justified. If we refrain from rewriting the past in this way, many scientific decisions look more reasonable. But, as Purtill notes ([62], pp. 57–58), scientists who were *sure* of their favored theory before some evidence for it was in were not justified, although they may well have been effective promoters of the theory.

Insensitivity to available counterinstances can also be a serious issue; some instances of (5) and (6) can thus be serious. Adamant insensitivity to available counterinstances conflicts with the control of hypotheses through their reference to publicly ascertainable facts of observation and experimentation. To adhere to a theory in the face of available counter-instances is to protect it from the test of experience. It would allow us to believe whatever we wish to believe, come what may. Such an ascription of insensitivity is, however, ambiguous, as Scheffler notes ([73], pp. 86–88). It could be taken to mean that a theory is retained along with statements acknowledged to be actual counterinstances to it. If this were so, it would amount to the charge that scientists are willing to tolerate logical inconsistencies; there is, however, little, if any, evidence for the plausibility of such a charge. But it could also be construed in a weaker sense. The weaker charge would claim that *prima facie* counterinstances to a favored theory are not accepted as actual counterinstances. They would be held to be due to an error as to the correct ascertainment of the initial conditions of the experiment or auxiliary assumptions rather than to an error of the theory itself. Even Hempel, a chief stalking-horse of radical meaning variance theorists, notes that in principle it would always be possible to retain a theory in the face of adverse test results, provided that we are willing to make sufficiently radical revisions among our auxiliary assumptions ([32], p. 28). Such practice would not amount to scientists' tolerating inconsistency; they would rather be preserving consistency in a biased manner so as to always protect their favored theories from overthrow. But such practice, if carried too far, *will* conflict with the scientific goals of simplicity and comprehensiveness. If more and more assumptions and hypotheses are specifically introduced to preserve and reconcile a partic-ular theory with the new observational evidence that becomes available, the resulting total system can become very complex and *ad hoc*. If so, it will conflict with the scientific goal of providing a systematically simple and comprehensive account of experience.

As the logician knows well, not all decisions and arguments are justified. We can grant for the sake of argument that there have been many too-early enthusiasts for scientific theories which only later proved to be successful. And we can grant that there were many scientists who wouldn't accept novel, though successful, theories and that there were even some who never accepted them and died not accepting them. We can, that is, grant that there exist pernicious instances of (5)–(7). But this, as Purtill ([62], p. 58) and Scheffler ([73], pp. 71–73) note, is in good part to point out that the logic of the sciences, like other areas of logic, is a normative, and not merely a descriptive, study. We will say more on this last point in Section VID where we will consider the counterclaim.

Construed correctly, some instances of (5)–(7) would thus provide Kuhn with historical evidence that scientists do not always act justifiably.[14] Yet, *if* Purtill's interpretation is correct and these *are* Kuhn's reasons they would not warrant his theses that in paradigm disputes "neither proof nor error is at issue" ([42], p. 150) and that the "transfer of allegiance from paradigm to paradigm is a conversion experience that cannot be forced" ([42], p. 150). Nor would they warrant his rejection of the "past" view that his historical evidence shows merely that "scientists being only human cannot always admit their errors" ([42], p. 150). Indeed, I have suggested that the past view is the correct view. I have argued that on Purtill's interpretation Kuhn's premises would not warrant his conclusion. Assume scientists do not always change or retain paradigms on rationally justifiable grounds. This simply would not, as Purtill notes ([62], p. 58), warrant the conclusion that there are no rationally justifiable grounds for paradigm change or that no standards – such as testability or falsifiability – are available to evaluate the choices men in fact make. Even if it could be shown that scientists *never* in fact employed rational grounds for paradigm change it would not *prove* the non-existence of such grounds; yet it might well be said to render their existence unlikely. Kuhn, however, (on this interpretation) would be far from establishing so sweeping a generalization. At best, *some* of his historical cases show that *sometimes* scientists engage in paradigm disputes without the employment of rational grounds.

Kuhn claims that:

Because scientists are reasonable men one or another argument will ultimately convince many of them. But there is no single argument that can or should persuade them
all. ([42], p. 157)

But is Kuhn's claim that scientists are reasonable men itself reasonable?
Is, as Purtill asks ([62], p. 58), making an important decision for reasons
that are supposed to be *only* persuasive acting as a reasonable man?
Perhaps so, if falsification and testability are not possible. But that they
are not possible is, I have argued, just what Kuhn, on the above interpretation of his position, has failed to show.

Further, as Scheffler notes ([73], pp. 88-89), Kuhn himself reintroduces
these very notions as relevant to the critical evaluation of competing
scientific theories. He opposes, as we have seen, the notion of falsification
along with its critically evaluative function. Yet he introduces the concepts of anomaly and crisis, along with a predictive criterion (cf., e.g.
[42], pp. 66–68, 96). And these have a similar function in the positive part
of his own account. Not every falsifying instance of a theory need produce
an anomaly or crisis. But, it would seem that every anomaly, crisis, or
predictive failure is at least a falsifying instance.

VI

There is a fifth methodologically undesirable consequence of the doctrine
of radical meaning variance. If it were true then there would be no sense
left to the notion of a rational *progression* of scientific viewpoints from
age to age.

A. Our conclusions in Chapter 1 and Sections II–V of this Chapter would
seem to imply this result. It follows from the doctrines espoused by
radical meaning variance theorists that two different theories could
neither agree nor disagree (II), that each scientist is effectively isolated
within his own and unique system of meanings (III) as well as his own
universe of observed things (Chapter 1), that no theory can be falsified or
tested (V), that scientists could not learn new and different scientific
theories (IV). Thus, it would seem to also follow that there is no sense to
the customary notion of the rational progression of scientific viewpoints
from age to age; the entire scientific enterprise would in principle be

subjective.[15] If contemporary theoretical alternatives cannot be compared and evaluated with respect to their factual accuracy or meaning, neither can succeeding theoretical alternatives; nor, for the same reasons, could reduction of one theory to another take place.

Justificatory processes of science would be non-neutral; they would have to be held to fail of objectivity. Personal factors would permeate not only the genesis of scientific theories but also their evaluation. The fundamental distinction between invention, generation, or discovery, and evaluation or justification, as urged e.g. by Einstein ([13], pp. 7, 13) and Reichenbach ([67], pp. 4–7) would have to be denied.

The radical meaning variance theorists' interpretation of science would thus seem to eventuate in relativism. It would become impossible, as a consequence of their views, to compare any two different scientific theories or to choose between them on any but subjective grounds. Scientific transitions would become complete and incommensurable replacements. Scientific change would have to be viewed as something similar to what happens when I, for subjective, aesthetic, or personal reasons, change my major area of specialization from Quantum Mechanics to Keynesian Economics or Sociological Theory; the latter changes would not, in an ordinary sense, be termed either 'progress' or 'regression' and neither would scientific change within the same field, given the radical meaning variance view. Meanings, whether of observational or theoretical terms, are theory dependent and are held to be radically and incommensurably different for different theories. If this is so, they can neither be compared directly nor indirectly with one another;[16] in particular, one could not be adjudged as better than the other. Being incomparable, there is no basis for choosing between them. Choice must be made arbitrarily. Therefore, when scientists choose one theory over another we could not claim, in any ordinary sense, that this choice constituted *progress*. In particular, we could not claim that the scientific community's choice of Einstein's theory over Newton's theory and of Kepler's theory over Brahe's theory constituted progress.

It will perhaps be instructive to see how Kuhn's views in particular give rise to such a conclusion. The problem arises because Kuhn gives no reasons consistent with the rest of his position, which could serve as grounds for accepting one paradigm as better or more acceptable than another. Given his general position one could not say, in any ordinary sense,[17] that *progress* is made when one paradigm is replaced, through a scientific revolution, by another. Why? Different paradigms radically disagree as to what are the facts, the problems faced, and the standards, which a successful theory must meet. A paradigm change brings about "changes in the standards governing permissible problems, concepts, and explanations" ([42], p. 105). The decisions of a scientific community to adopt a new and different paradigm could not, therefore, be based on good reasons of any kind, factual, or otherwise. Kuhn's position indeed tends toward the conclusion that the replacement of one paradigm by another is not cumulative, but is mere replacement, mere change. He maintains, if the interpretation of Kuhn proposed by Shapere ([76], p. 391; [78], p. 67), Purtill ([62], pp. 53–54), and Scheffler ([73], p. 82) is correct, that facts, problems, concepts, and standards are each defined by the paradigm and are each radically and incommensurably different for different paradigms. If this is so two different paradigms could not be judged according to their ability to solve the same problems, deal with the same facts or concepts, or meet the same standards, for all of these are radically different for different paradigms.

Such a conclusion is, however, inconsistent with the positive part of Kuhn's methodology. In discussing how paradigm disputes are finally resolved ([42], p. 168), Kuhn draws attention to "two all-important conditions" which a new and successful paradigm will satisfy. First, it will "resolve some outstanding and generally recognized problem that can be met in no other way." And second, it will "preserve a relatively large part of the concrete problem-solving ability that has accrued to science through its predecessors"; it will "preserve a great deal of the most concrete parts of past achievement." However, neither of these "all-important conditions" could be met if the rest of Kuhn's interpretation were correct. As Scheffler has correctly noted:

Such conditions of evaluation contradict the main thesis appealing to the history of science ... namely, that paradigm change in science is not generally subject to deliberation and critical assessment. ([73], p. 89; cf. Kuhn, e.g. [42], p. 150)

The consequences that I have drawn in this section from the general radical meaning variance position are, from a methodological point of view, surely undesirable, if not false. Objectivity is, all things being equal, preferable to subjectivity; striving for progress would be preferable to striving for non-progress (mere change) unless, perhaps, progress is demonstrably unattainable. In Chapter 2, however, I argued that the radical meaning variance position was in fact not demonstrated; the arguments and criteria for it failed. And in this Section I have just argued that this position would preclude scientific progress. Thus, scientific progress is not demonstrably unattainable via the arguments for the radical meaning variance position. Kuhn, however, adduces other arguments which purport to more or less directly demonstrate this unattainability. I shall examine these in the next section (VIB). Following this I will examine Toulmin's early attempt (1961) to redefine the concept of scientific progress with regard to how far paradigms take us and how theoretically fruitful they are (VIC). Next (VID) I will consider Toulmin's recent (1967) way of redefining this concept along with his claim that a scientific theory should be considered better than another only because it is accepted.

B. Science, for Kuhn, could not be a cumulative enterprise. Scientific change could not constitute *progress*, in any ordinary sense, since each time a paradigm change occurs, what science is trying to do changes ([42], pp. 93, 108–109, 139–140, 149–151). In short, Kuhn feels that paradigm change could not be justified, for all justifications involve paradigms. Let us examine this carefully. Supported as it is by a wealth of historical examples it tends to look impressive. But, for Kuhn, this depends on a particular interpretation of his evidence which may be questioned.

Let us assume that all justifications of paradigm change involve paradigms. It is my claim that this assumption does not commit us to holding that no paradigm change can be justified. In any scientific change, those features of paradigms which presumably give meaning to the terms may not be the features at issue in the change. And this may be true even for those changes which Kuhn would call 'revolutions'. Consider the much discussed example of the scientific revolution of classical mechanics to relativistic mechanics. In both theories *mass* can be characterized as the ratio of force to acceleration. In Hamiltonian formulation *linear momentum* can, in each theory, be characterized as a certain space integral of the Hamiltonian. To say that these are not parts of the meanings of 'mass' and 'linear momentum' respectively would be arbitrary and to beg the question. To say that *only* those characterizations of the terms which do change should be allowed to count as "part of the meaning" of the term is unwarranted and again to beg the question. Those very features or functions which do not change in the use of a term may be important features and functions; they may, for some purposes, prove to be the very ones that are of central importance in comparing different theories.[18] And some important functions and uses of scientific terms have remained invariant with respect to scientific revolutions, as our above two examples illustrate. In other words, certain features are in a sense shared by competing paradigms and are common to both of them. And this is part of what makes rational justification of paradigm change possible. It is a commonplace that rational argument cannot take place unless some things are held in common: some shared premises, some agreement as to what counts as an argument, some evaluative standards. An application of this to paradigm change and debate suggests that for there to be justification of paradigm change there must be some agreement between or about competing paradigms. My above two examples ('mass' and 'linear momentum') show that there has in fact been significant agreement between competing paradigms. They suggest that scientists engaged in a paradigm dispute may be able to talk to, not through, one another; they suggest that the logical status of such debates and disputes is not always "necessarily circular" or "only that of persuasion" ([42], p. 93); they suggest that the two theories are neither radically different in meaning nor incommensurable, but rather, commensurable, and perhaps comparable through appeal to shared meaning. That revolutions change much less than they seem to on the surface is a historical truism; and this truism is relevant to questions of paradigm change.

Kuhn argues that paradigm change is so radical and complete that scientists cannot communicate or argue rationally across the gap. I would suggest that the gap is much exaggerated, as it always is in revolutionary changes. In short, paradigm change might, I think, plausibly be viewed for evaluative purposes as a deliberative process which can occur because of features shared by competing paradigms.

C. There is a second sort of objection to my claim in VIA that if the radical meaning variance thesis were true then scientific change could not constitute progress. Toulmin [83] claims, in effect, that there is a special sense of 'progress' and 'cumulative' in which the radical meaning variance position does not entail that scientific change is non-cumulative or non-progressive. After accepting the radical meaning variance thesis as to both observational and theoretical terms ([83], pp. 57, 95), he exhibits some sensitivity to the problem we raised in VIA:

...how do we know which presuppositions to adopt? Certainly, explanatory paradigms and ideals of natural order are not 'true' or 'false' in any naive sense. Rather, they 'take us further (or less far)' and are theoretically more or less 'fruitful.' ([83], p. 57)

Given Toulmin's radical meaning variance position, it would be doubtful, as Shapere has pointed out ([77], pp. 18–19), whether there are shared ideals or standards which are invariant with respect to scientific revolutions. Therefore it would be doubtful whether different paradigms could be judged to be "more or less fruitful" in accomplishing common tasks or fulfilling common purposes. For the same reason it would be doubtful whether there were common goals that one paradigm could "take us further" toward than another. It would be doubtful whether one set of ideals of natural order, one set of "basic" presuppositions, or one paradigm, could be better than another in the sense that it would enable us to more fully meet the same standards, or deal more effectively with the same problems or with the same facts, or more systematically and simply arrange the same or similar concepts and categories. This is because each of these would presumably differ radically from paradigm to paradigm given (say) Toulmin's radical meaning (and observational) variance position. Indeed, Toulmin claims that men "who accept different ideals and paradigms have really no common theoretical terms" in which to even "discuss their problems fruitfully" ([83], p. 57).[19]

Like Kuhn and Feyerabend, Toulmin here offers us little in the way of a viable specification or clarification of the sense in which one theory can be judged better or more acceptable than another, given his radical meaning variance position. The question of how theories can be judged against one another and how the replacement of one theory by another can be said to constitute "progress" or an "advance" remains unanswered. We must, therefore, conclude that Toulmin's answer to the difficulty which we posed in VIA is unsuccessful. The difficulty is a major problem and remains.

D. There is a third sort of answer to the problem we posed in VIA, viz. to the problem of how one could, given the radical meaning variance doctrine, say that "progress" results from scientific revolutions. It is similar to the reply we have just discussed (VIC) in that it accepts both radical meaning variance and the consequence that there can be no cumulative or progressive growth of scientific knowledge in any standard sense; and as with the previous reply, it does not conclude that the radical meaning variance thesis is methodologically undesirable. It concludes rather "that scientific progress is not what we had taken it to be" ([42], p. 169). This reply maintains that "scientific progress" can, and should, be redefined in a non-standard, or non-ordinary sense. Succinctly put this reply claims that a theory should not be accepted by a scientific group because it is better than any alternatives; on the contrary, it should be called 'better' only because it is accepted. The decisions of a scientific group to adopt a new paradigm should not be based on any externally good reasons, factual or otherwise; quite the contrary, what counts as a good reason should be determined solely by the decision. The theory that the scientific community adopts should always be regarded as the theory which constitutes scientific progress. Kuhn would say that,

As in political revolutions, so in paradigm choice – there is no standard higher than the assent of the relevant community. ([42], p. 93)

The sort of reply that I have reconstructed of course brings with it its own methodology of science, a purely descriptive methodology. Such a reply and accompanying methodology were first suggested by Kuhn ([42], pp. 93, 161, 165, 169). And recently they have been explicitly adduced by Toulmin [84] and [85] as constructive answers to the criticisms he raises with respect to the views of Collingwood and Kuhn concerning conceptual revolutions in science; his criticisms are similar to some of those I have raised in this Chapter in connection with the radical meaning variance doctrine; however, they are, in my opinion, equally applicable to Toulmin's earlier views [82] and [83].

Toulmin feels that his view will lead to a view of conceptual change which preserves the merits of the analyses of Collingwood and Kuhn while avoiding their shortcomings ([84], pp. 86–90).[20] He thinks that in generalizing about the merits of rival scientific theories, philosophers of science should describe the selection criteria that specific professional groups in fact use to guide their choices between conceptual alternatives ([84], pp. 87–88). Further, he claims philosophers of science should *only* describe these criteria ([84], p. 89). This is a distinct advantage since then "...philosophers can no longer dictate the principles to which scientists *ought* to conform..." ([84], p. 89). His remarks here come down to the following thesis:

(T)   In generalizing about the merits of rival scientific theories philosophers of science should use only a descriptive methodology.

(T) is, in my opinion, untenable. For in holding it philosophers of science would be methodologically prescribing that philosophers of science use only a descriptive methodology. And this activity itself – the *prescribing* – is not an activity of describing. Hence, in virtue of holding (T) a philosopher of science would be violating (T). In this sense then, (T) seems an untenable thesis.

However, one might reply as follows: The person who holds (T) is admittedly prescribing that philosophers of science be only descriptive. But holding (T) is a different activity (call it 'philosophy', not *'philosophy of science'*) than (T) is talking *about*. (T) is talking about the philosophy of science, not about philosophy in general. Thus, (T) is tenable.

But reinterpreting Toulmin in this way just won't work. Why, and for what reasons, is description appropriate for judging rival scientific theories, and prescription appropriate for judging rival philosophies of science? Toulmin's above remarks do indeed entail that we be prescriptive in judging rival theories of the philosophy of science; we *should* consider which of them is the more descriptive insofar as they talk about science. But why be prescriptive in the one case and descriptive in the other? It would be incumbent on Toulmin to give reasons why prescription is appropriate in judging the merits of different philosophies of science, but not in judging the merits of different scientific theories. What is the difference between these two domains that makes this so? Is it because hard-and-fast and "*non*-theory-laden" observational facts are relevant to science, but not to the philosophy of science? If Toulmin answered 'yes' to the latter question, as he would seem to have to, he would be dangerously close to the "logical empiricism" he wishes to deny ([84], pp. 76, 90).

Further, how could it even be *possible* for one to generalize about the *merits* of anything by using only a descriptive methodology? When Toulmin says,

In generalizing about the *merits* (my italics) of rival scientific theories... we should concern ourselves with the selection criteria which *in fact* (his italics) guide the choices men make.... ([84], p. 88)

one could always reply to any particular case, 'But *ought* the men to have made these choices?' To this question Toulmin can have no answer. But it must be answered if we are to *appraise* ([84], p. 90) different scientific theories and generalize about their *merits*.

The situation is, however, still worse. Adherence to Toulmin's thesis would tend to prolong the acceptance of unsatisfactory scientific hypotheses if everyone happened to agree with them. If this thesis had been accepted by scientists and philosophers during the middle ages, I doubt if modern science ever would have arisen, for "modern" science was advocated by only a small minority of thinkers during this time. If scientists had accepted Toulmin's account of the nature of their own activity they therefore would have been hindered, not enlightened. And this just seems to show that the hypothesis that the reasonableness of scientific hypotheses is essentially dependent on people's acceptance is itself an unreasonable hypothesis.

Galileo *did* in fact choose to repudiate his scientific theory while under pressure from the inquisition. But is *this* repudiation relevant for judgments about the intellectual merits of his theory? Certainly not. Is conceptual change to be appraised *solely* by looking at the reasons the men who made the change *in fact* gave? Or is it to be judged by looking at these reasons in light of some norm or standard? If the latter is the more reasonable position as I have intimated, then what is needed in judging about science is precisely what Toulmin *denies*. We do need *in addition to* historical studies, some kind of "abstract and quasi-mathematical schematization of 'inductive logic'" ([84], p. 76). We do need some "timeless and a-historical standards" ([84], p. 76). And Toulmin has, in his inordinate dissatisfaction with "logical empiricism", denied these needs.

I therefore conclude that Toulmin's recent reply, which was previously suggested by Kuhn, has failed. It has not provided a viable sense for a redefinition of scientific progress. My objections raised in this Chapter, and particularly that of VIA, thus remain.

VII

The sixth methodologically undesirable consequence of the doctrine of radical meaning variance is perhaps the worst. The doctrine is demonstrably untenable because of a self-referential problem.

If my conclusions in Sections II–VI are correct one must conclude that scientific objectivity is, for the most part, a myth if the doctrine of radical meaning variance is correct. Some writers at times seem to recognize and, surprisingly, accept this conclusion. Kuhn, Feyerabend, Polanyi, and Whorf have each expressed themselves in this way. Kuhn writes:

We may, to be more precise, have to relinquish the notion, explicit or implicit, that changes of paradigms carry scientists and those who learn from them closer and closer to the truth. ([42], p. 169)

He says that he

... would argue, rather, that in these matters neither proof nor error is at issue. ([42], p. 150)

On the basis of his interpretation of the history of science, Polanyi, also, feels that in a strong sense scientific knowledge is not objective ([60], p. 16). On the basis of his study of widely differing languages Whorf makes a similar claim:

The relativity of all conceptual systems, ours included, and their dependence upon language stand revealed. ([87], pp. 214–215)

Feyerabend, too, perhaps because of a felt lack of success with his previous attempts to objectively compare different theories, has very recently come around to such a view:

... an enterprise whose human character can be seen by all is preferable to one that looks 'objective', and impervious to human actions and wishes. The sciences, after all, are our own creation, including all the severe standards they seem to impose upon us. It is good to be constantly reminded of this fact. ... What better reminder is there than the realization that ... the choice of our basic cosmology may become a matter of taste? ([19], p. 228)

Such are some consequences of the radical meaning variance position. As Scheffler notes ([73], p. 18), "adoption of a new scientific theory" would be "an intuitive" "mystical", or merely, aesthetic affair; it would be a matter primarily for "psychological description" rather than for "objective" logical and methodological justification or evaluation in the usual senses of these terms.

Such wholesale rejection of objectivity, however, is problematic. Objectivity is presupposed by any statement which purports to make a cognitive claim. To put forth any such claim in earnest involves a presuppositional commitment to the view that the claim has an objective truth value. Consider the particular claim that no claim is objective. It is itself put forth as an objective truth. If, however, it is true then by its very statement it must be false.

There is, as Scheffler has noted ([73], pp. 21–22), problematic self-contradictoriness: (a) in the earnest effort to communicate to others that communication is impossible; (b) in appeal to neutral facts about observation in order to deny that neutral facts of observation exist; (c) in arguing from the given realities of the history of science "to the conclusion that reality is not discovered but made by the scientist". "To accept these claims implies that we are" – in the absence of objective standards – "free to assert what we will"; "hence, in particular", we are free "to reject them" ([73], p. 22). The statement that it is impossible for there to be statements whose meaning or truth is invariant from theory to theory is itself a statement in fact put forward as true and as having the same meaning no matter what different theories we may hold. If it were true we would therefore have it itself as an example of a statement whose truth and meaning stays the same no matter what different theories we may hold; thus, by its very statement, it would be false. Hence, it *cannot* be true.

When Whorf asserts that "the relativity of all conceptual systems, ours included, and their dependence upon language stand revealed", he is making a statement which he considers to be: (a) true; (b) part of his total conceptual system; and (c) determinate in meaning. If, however, this statement is true then, since it is expressed in English, it too must be relative to the English language. If this is so then, this particular meta-claim might, presumably, not hold in a different language and associated conceptual system. If it were false in some conceptual system, by its very statement, it would be true that there exists at least one conceptual system which is not relative and which is independent of language. But if this is true then Whorf's self-referential meta-claim is false, even when expressed in English. In short, assuming it true, we can deduce that it is false; it is, therefore, untenable.[21] Similarly, the claim that there is no objectivity itself purports to be objective. Hence, if it is true, it must be false. It is, therefore, an untenable claim.

One might object by reinterpreting the above consequences. Roughly, one might want the claim to be only that there can be no objectivity *in science*. There would be no self-referential problem in then, for example, appealing to neutral meta-facts which are *about* observation in order to deny that neutral facts *of* observation could exist.

Let us, therefore, examine the position that there is no objectivity *in science*. Consider the claim that the choice between only scientific theories is "a matter of taste" (Feyerabend) where "neither proof nor error is at issue" (Kuhn). Such a restricted claim although not self-referentially inconsistent like the more general claim, is, nevertheless, problematic. It leads to an unjustified dualism. On the one hand we are supposed to hold (8):

(8)   Science is a subjective enterprise (in the sense specified above, cf., e.g., Feyerabend and Kuhn) whose concepts and domain are theory-laden.

But, (8) is indeed put forward as the true, correct, proper, etc., way to view scientific transitions. Thus, on the other hand, in order to claim (8), and on the reinterpretation of these matters that we are considering, we are supposed to also hold (9):

(9)   The philosophy of science is an objective enterprise whose concepts and domain are not theory-laden.

People who adhere to (8) would, to avoid the self-referential problem, *have to* maintain that their own views of scientific change are uninfluenced by the fact that they are (say) Kuhnians; that is, they would have to hold (9). Assume they instead held that their own views of scientific change were so influenced, and were, therefore, as with scientific theories, also "a matter of taste" where "neither proof nor error is at issue"; in short, assume that they held that their own views of scientific change were subjective. Then they would have to hold that (say) Kuhn's view of scientific change is no more correct than (say) Hempel's "logical empiricist" account. This, however, neither Kuhn nor Feyerabend would want to do. Each believes "the logical empiricist" account to be defective and in error. And each believes his own account to be correct and to remedy the objective excesses of logical empiricism. The dualism, therefore, remains. The facts scientists deal with are infected by the particular theory employed. The facts which (say) Kuhnian philosophers of science deal with are unexplainedly uninfected by their meta-theory, even though this theory is far more *conceptual and theoretical* in nature than any scientific theory. What is the difference between these two domains that makes this so? Is it because hard-and-fast and "*non*-theory-laden" facts are relevant to the philosophy of science but not to science? If this were affirmed, radical meaning variance theorists would be dangerously close to a new sort of "logical empiricism" at a meta-level, one quite similar in structure to the old one they wish to eschew. Further, such a dualism would be at odds with a tendency towards precisely the reverse sort of dualism implied by (say) Kuhn's and Toulmin's advocation of a purely descriptive methodology for evaluating rival scientific theories (cf. section VID and [39], pp. 459–460). Our conclusion, therefore, is that the doctrine of radical meaning variance is either demonstrably untenable (for it gives rise to consequences which are self-referentially inconsistent) or else leads to an unjustified and problematic dualism with "neo-positivistic" aspects.

CONCLUSION

It is unreasonable to assume that scientific transitions force a radical change in the meanings of the terms employed which precludes comparisons of different theories through appeal to shared meaning.

NOTES

[1] As Kuhn notes: "[Brahe] himself had looked for parallax with his great new instruments. Since he found none, he felt forced to reject the earth's motion" ([41], p. 201). Further, "Brahe's skillful observations were even more important than his system in leading his contemporaries toward a new cosmology. They provided the essential basis for the work of Kepler, who converted Copernicus' innovation into the first really adequate solution of the problem of the planets" ([41], p. 206). In spite of Kepler's mysticism – which incidentally he *argued for* on metaphysical grounds, and *not* by a linguistic analysis of the meaning of 'sun' – he still sought confirmation for his heliocentrism through experiment. Thus, "without proper experiments, I conclude nothing" ([37], V, 224; cf. also I, 143). Interestingly enough, Hanson himself notes this: "They [Brahe and Kepler] met in Prague in 1600. Brahe was at this time working on his theory of the orbit of Mars, which Kepler found unsatisfactory, since Brahe's miscalculations amounted sometimes to a 5° *disagreement with observations* (this was not excessive for the geocentric theories then prevalent). Kepler adopted the heliocentric view" ([27], p. 73, my italics). Cf. again Kepler: "... after I had by unceasing toil through a long period of time, using the observations of Brahe, discovered the true distances of the orbits, at last, at last, the true relation... overcame by storm the shadows of my mind, with such fullness of agreement between my seventeen years' labor on the observations of Brahe and this present study of mine that I at first believed that I was dreaming..." (*Harmony of the World*, 1619).

[2] A similar objection also has been raised by Achinstein ([1], p. 499), Shapere ([78], p. 57), and Fine ([20], pp. 231–232). Feyerabend recently [18] has accepted the force of this objection and attempts to circumvent it by providing a new sense for the predicate 'incompatible' as applied to different theories. We will examine his answer in the next section.

[3] For some technical details cf. [26], pp. 1099–1103.

[4] For some technical details cf. [36], pp. 271–273.

[5] Three other replies, written earlier, to somewhat similar problems are to be found in [16], pp. 216–217. Each is admirably criticized by Shapere, [78], pp. 58–61.

[6] For some technical details on the concept of an isomorphism which suggests these results, cf.: [38], pp. 221–230; [57], pp. 90–92.

[7] I have recently discovered that Achinstein ([2], pp. 93–94) provides a surprisingly similar counterexample to Feyerabend's proposal by using Keynesian theory and quantum theory. Our respective remarks concerning Feyerabend's proposal, though independently arrived at, are quite similar and to the same point.

[8] Shapere ([75], esp. p. 145) has suggested that this claim itself may not be intelligible.

[9] This consequence of radical meaning variance has also been noted by Achinstein ([1], p. 499; [2], pp. 97–98) who presents an admirable discussion of the problem.

[10] Most writers within the logical empiricist tradition (e.g. Carnap, [8], and Hempel, [31]) would not be open to this argument. They would maintain that a set of "reduction sentences" determines the meaning of a term "only partially" ([31], p. 28). Theoretical concepts would be permitted to enter new general principles and laws in order to preserve "*openness of meaning*" for the technical terms of science ([31], pp. 28–29). Yet, prior to their entering into the new generalizations and laws, the terms are held to already possess (in virtue of "'operational' criteria of application ... expressible in terms of observables") some determinate "operational meaning" which remains invariant ([31], p. 29).

[11] Achinstein's, otherwise excellent, discussion ([2], pp. 97–98) fails to recognize this point.

[12] Just this, however, was not done when light bent when travelling near the sun.

[13] Central to his view is that T' is a more *efficient* alternative for purposes of testing T the more radically different in meaning T' is from T (cf., e.g., [15], pp. 7–8).

[14] However, as I argued above, this would occur less frequently than Kuhn would think.

[15] Feyerabend has very recently come to recognize this consequence, cf. [19]. We will discuss his remarks on this point in Section VII.

[16] Neither Kuhn nor Feyerabend has succeeded in providing any extra-theoretical or cross-theoretical basis or approach which is both consistent with their radical meaning variance position and which would also constitute a means by which incommensurable theories could be compared in an indirect way (cf. IIB, end of V). In this regard, however, we will also examine Toulmin's attempts in VIC and VID.

[17] I will soon examine (VIC, VID) two non-ordinary senses; one is suggested by Toulmin and another is hinted at by Kuhn and later explicitly advocated by Toulmin.

[18] For some good illustrations and further substantiation of this point cf.: Purtill, [62], pp. 55–56; Fine, [20], pp. 237–239; Shapere, [78], pp. 71–81; and Achinstein, [1], pp. 503–508.

[19] We would be left with only the rather weak notion that one theory is better than another if it deals more effectively with its *own* facts. This alone does not do full justice to the concept of scientific progress. It would not be able to distinguish (a) from (b):

(a)   Progress was made when classical mechanics was rejected and quantum mechanics was accepted.

(b)   Progress was made when Freudian Psychology was rejected and quantum mechanics was accepted.

The relations "is a rival of", "is an alternative to", "is a competitor of", which are of prime importance for the understanding of (a) and for the understanding of the difference between (a) and (b), would be lost.

[20] The following discussion of Toulmin's view is based on my article, [39].

[21] The above points up some conceptual difficulties with Whorf's hypothesis. There is, however, empirical evidence which also disconfirms his hypothesis, cf. M. Cole *et al.*, [9].

CHAPTER 4

THE COMPARABILITY OF SCIENTIFIC THEORIES

I have argued in Chapter 2 that the proposed arguments and criteria for radical meaning variance have failed. And in Chapter 3 I have argued that this thesis is in itself both incorrect and methodologically undesirable. In this Chapter I wish to discuss some of the problems which have motivated radical meaning variance theorists and point up some distinct virtues and advantages of their account as against a more traditional logical empiricist approach. Using my conclusions in Chapters 1 to 3, I then wish to suggest an alternative account.

I

Radical meaning variance theorists have been dissatisfied with the logical empiricist tradition. They think that an entirely new *approach* to the philosophy of science is required.

Radical meaning variance theorists have, in particular, been dissatisfied with the general logical empiricist attempts to approach the philosophy of science by application of the techniques of mathematical logic. Logical empiricists have treated scientific theories as semantically interpreted formal systems with precise rules of interpretation. However, in their concentration on the technical problems of logic arising from this approach, they have tended to lose close contact with science as it is in fact practiced. Logical empiricists often become involved with logical details. They usually analyze only thoroughly developed scientific theories. They seldom analyze or examine the historical development of actual science. They have therefore become open to the criticism of being too rationalistic, despite their professed empiricism. With some justification they have been charged with failing to attend to facts which constitute the subject matter of the philosophy of science.

Feyerabend, Hanson, Kuhn, and Toulmin have each sought to remedy such defects of the logical empiricist approach. And they have met with at least partial, and often illuminating, success, my objections in Chapters 2 and 3 notwithstanding. Most of these writers have been influenced by Wittgenstein's warning that a great many functions of language must be ignored when it is viewed simply as a calculus without regard to its actual use. They have profitably viewed scientific theories in light of this warning. And a noteworthy merit of the new approach of these writers is their detailed concern with the history of science. The logical empiricist tradition has tended to view the history of science as virtually irrelevant to the philosophy of science. It tended to look on the history of science as a chronological record of the slow removal of superstition, prejudice, and other obstacles to scientific progress. And *when* logical empiricists *have* concerned themselves with historical matters, they have been almost exclusively concerned with completely developed theories, as Hanson would say, with "the finished Research Report". Such a restricted concern, however, often leads them to ignore the ways and rationale by means of which incomplete theories in fact develop. Hanson, for example, frequently stressed this point in his work on the notion of a "Logic of Discovery". And in their new approach, philosophers like Hanson, Feyerabend, Kuhn, and Toulmin have brought out features of science which clearly conflict with traditional forms of logical empiricism. One of the interesting results of their approach is that many scientific theories – e.g. Aristotelian and medieval mechanics, the phlogiston and caloric theories – have been discovered to contain far less simple-minded error and superstition than was initially thought.

The radical meaning variance thesis which each of these writers would espouse is motivated by, and intended to account for, the fact that scientists holding different theories engage in different research projects. And this fact is, I think, undeniable. As Hanson succinctly notes "the difference in their conceptual organization guarantees that in their future research they will not continue to have the same problems" ([27], p. 118). Continued attempts to account for and understand the underlying reasons behind such facts from the history of science should illuminate much of the rationale of scientific transitions and scientific activity in general.

A good part of what has motivated the radical meaning variance thesis has also been a reaction to the logical empiricist tradition's insistence on a meaning-invariant observational language which different theories are held to share. In part also the new approach and thesis has been due to, and intended to account for, the fact that after scientific "revolutions", scientists in fact do use scientific terms in new ways. In their heavy reliance on the equation of meaning with use, this fact becomes crucial for many radical meaning variance theorists. And they have argued, as we have seen, that scientific "revolutions" force scientific terms to change radically and "incommensurably" in meaning.

But, although they reject the notion of a meaning-invariant observational language, radical meaning variance theorists meet with little success (cf. Chapter 3) in solving problems this device was in part designed for, how one theory could be adjudged as compatible or incompatible with, better than, more acceptable than, a rival of, or in competition with, another. Logical empiricists attempted this by recourse to the shared *meanings* of the terms employed by each theory. They resorted in particular to the presumably neutral and meaning-invariant observational language employed by different theories. Radical meaning variance theorists have rejected a neutral and meaning-invariant observational language. They have claimed that different theories are incommensurable in meaning. They therefore cannot appeal to the meanings of the terms employed by each theory in order to make the above comparisons. And their attempts at comparison without this appeal have failed, as we have seen in Chapter 3 (IIB, VC, VD). In contemporary philosophy of science therefore the problem of the comparability of different theories has increased in seriousness and importance.

As Shapere stresses ([79], pp. 117–118), the positivistic tradition within twentieth-century philosophy of science has not only utilized the notion of a meaning-invariant observational language in order to analyze the meaning and acceptability of *single* theories. It also uses this device to provide a basis by which *different* theories can be said to be "in competition" and as a basis by which one theory can be chosen as "better than" its competitors. The positivistic tradition would hold that because two different theories deal with the same observations and have the same observational vocabularies the ground for such comparisons is the common meanings of the terms employed by each theory. Logical empiricists have usually held that the meanings of the theoretical terms of any theory are determined, at least partially, by the meanings of the observation terms with which they are somehow correlated (through operational definitions, reduction sentences, etc.). They have held that observational terms have a stable meaning in reference to which the theoretical terms of different theories receive their meaning. Logical empiricists would therefore maintain that different theories can be compared through appeal to the meanings of the terms they employ since these terms are correlated with a theory-neutral and meaning-invariant observational vocabulary. And one theory would be judged "more acceptable than" a "rival" if, for example, it were better confirmed than the other theory *and* if each theory employed the same observational vocabulary.[1] For positivists the solution of any problems concerning the comparison of different theories – e.g., the relative merits of competing theories or even how they could be said to be in competition at all – would follow as a matter of course once the exact details of the basis for interpreting and evaluating single theories themselves were developed. However, in the last few years the problem of the comparability of different theories has increased in seriousness and importance. It is now a central problem in the philosophy of science. This has been due primarily to two groups of factors: (a) criticisms of the claim that successive scientific theories each employ a completely meaning-invariant observational language; (b) the difficulties encountered by radical meaning variance theorists who in their denial of the positivistic approach attempt to compare different theories without appeal to the meanings of their respective terms (cf. Chapter 3).

Radical meaning variance theorists have also exposed shortcomings of usual formulations of the distinction between theoretical and observational terms. Such writers do indicate ways in which the meanings of "observation terms" may involve a theoretical background. In short, the logical empiricist tradition has perhaps overemphasized the *invariance* of meaning in scientific change. Successive scientific theories were held to employ a completely meaning-invariant observational language. Actual science, however, does not exactly proceed in this way. After scientific "revolutions", scientists do use scientific terms in some new ways. For these reasons and because such terms enter into some new and different relationships within different theories, one *might* sometimes wish to say that the terms employed by successive theories have changed meaning to *some* degree. But in order to note this it is not necessary to go to the opposite extreme from positivism. One need not maintain with radical meaning variance theorists that competing or rival theories employ terms whose respective meaning is *incommensurable* or *radically* different. Feyerabend and Kuhn, for example, have each overemphasized the *variance* of meaning in scientific change at least as much as logical empiricists have over-emphasized the invariance of meaning in such change. Science, however, does not in fact proceed in this way either. To say it does is, as Shapere notes ([79], p. 121), to overlook the fact that different theories and different usages of the same terms often exhibit significant similarities and continuity in the development of science.

In previous Chapters I have not denied that after scientific "revolutions", scientists do use scientific terms in some new ways. But, I have suggested that this is primarily because they believe that new properties, relations, and rules are applicable to experience and also because they perhaps believe that scientific terms enter into new essential relations. I have urged that not all of these new beliefs can be said to have changed the meanings of the terms employed. I have also suggested that even if there has been a change of meaning it need not be, and in fact usually has not been, radical, "incommensurable", or complete. I have suggested that there can be, and usually is, sufficient similarity of meaning of the terms em-ployed before and after scientific revolutions so as to permit different types of comparisons between the old theory and its successor. Com-parisons of different theories are possible since it is possible for them to use terms with shared meanings. An old theory and its successor need not be, and usually are not, incomparable in all of the ways the radical meaning variance thesis would force them to be (cf. Chapter 3).

II

I have suggested that comparisons of different theories are *possible* since it is *possible* for there to be some shared meaning of the terms in-volve. Further, I should like to urge that comparisons of different theories are *in fact* made through appeal to shared principles and meaning at *both* a "first" and "second" level. By shared meaning at a first level I mean shared extension of terms employed by rival theories.[2] And by shared principles and meaning at a second level I mean shared *regulative* prin-ciples which scientists require of successful theories and which guide their choice among alternative theories. I will discuss the nature and function of second level sharing in Section IV. In the present section I wish to examine first level invariance and argue that it usually occurs in scientific transitions. In the next Section (III) I will sketch the outlines of an account of the nature of observation in science; it should explain how and why first level invariance usually occurs, as well as provide additional support for the claim that it does. Taken together, first and second level invariance will enable us to compare theories in several ways, e.g., with respect to the relation "is a better competitor than".

A. In Chapter One I argued that the arguments for radical observational variance failed and that this thesis itself is implausible. This provided us with some justification for assuming that, in a non-trivial sense, observation is neutral with respect to alternate scientific theories. If we were correct then there does exist some aspect – the extensional aspect – of meaning which is usually invariant with respect to scientific transitions. Extensional invariance is part of the basis which permits cross-theoretical comparisons. If some observation is invariant then some extension is invariant. And a significant part of the meaning of a term is its extension. There are several reasons for this.

The existence of things referred to by some of our terms is presupposed, for semantic purposes, by any theory we might adopt. Whether the existence of senses or intensions is an indispensable assumption of any theory is, however, questionable (cf. [64], pp. 47–48; and [73], pp. 54–58).

Finally it is important to note, as Scheffler does, that "for the [semantic] purposes of [logic], mathematics and science, it is sameness of reference [or overlap of extension] that is of interest rather than synonymy" ([73], p. 57). Quine ([64], pp. 47–48, 132), Goodman ([23], Chapter 1; [24]), and Scheffler ([73], pp. 54–58, 62) each stress this point well. They urge that the notion of definition "in mathematical and scientific contexts, hinges not on synonymies [or shared intensions] but, at most, on referential equivalences" ([73], p. 57). Alterations of meaning in a valid deduction that leave referential values intact are irrelevant to the preservation of truth (cf. [73], p. 57).

These considerations suggest the usefulness of regarding extension as a first step toward an account of meaning. In any case, the extensional aspect of meaning is generally regarded as *a* significant aspect of meaning. Chapter 1 provided us with justification for assuming that, in a non-trivial sense, some observation, and therefore some extensional meaning, is neutral or invariant with respect to alternative theories.

B. Although we have not yet, in this Chapter, argued directly for first level invariance in actual scientific cases, the reader may have already felt the following objection: The purpose of establishing observational invariance is to establish meaning invariance. But observational invariance itself can be established only by presupposing meaning invariance. Thus, meaning invariance can be established only by presupposing meaning invariance. And this is viciously circular.[3]

This objection can, perhaps, be answered. Meaning (extensional) invariance between theories T_1 and T_2 indeed presupposes observational invariance between T_1 and T_2. But in order to establish observational invariance between T_1 and T_2 we need not presuppose meaning invariance *between T_1 and T_2*. We could *conceivably* describe the neutral observations, if they exist, by use of some third T_3. A fragment of ordinary language, or a suitable meta-language, might conceivably serve this purpose.

The issue here is perhaps compressed into this question – Is there such a meta-language? If one has a theory committed to the existence of a first time then one would clearly have an ontology that is *in this respect* different from an ontology of a theory committed to an infinite past. There would here be *some* ontological and extensional variance. In such examples there admittedly is no neutral meta-language which could describe and assist in establishing a co-extensive domain or ontology, for complete ontological or extensional invariance would simply be lacking. There would not exist a meta-language which could describe the objects and ontologies dealt with by such theories in a way which would allow us to establish the complete neutrality of these objects. This is simply because there is not in these cases complete object neutrality: what does not exist (co-extensive domains or ontologies) cannot be described.

Such examples, however, do not really militate against our position. We are concerned with whether there is *some* non-trivial object invariance. This involves us in inquiring into whether there can be and is a meta-language whose terms are independent enough of the particular theoretical commitments of two radically different physical theories and object languages so that these terms can serve to describe significant neutral objects (i.e. non-trivial objects in the set-theoretic intersection of the two ontologies). These concerns give rise to two questions: (a) Historically speaking *was* there such a neutral meta-language available to holders of rival physical theories? (b) Can and do *we* possess such a neutral meta-language (perhaps a fragment of ordinary or quasi-formalized English?) with which to describe and discuss significant common objects from the ontologies of rival physical theories of the past?

Radical meaning variance theorists would presumably answer both questions in the negative. With regard to scientific transitions they claim (cf. Chapter 1 and 2) that there does not exist significant ontological or extensional overlap; and what did not and does not exist (non-trivial ontological or extensional overlap and common objects) obviously can't be described either by anyone in the past or by us in the twentieth century.

On our approach, however, affirmative answers will sometimes be possible to both questions. There are not, in my opinion, any tenable or coherent *philosophical* reasons why (a) would *have to be* answerable in the negative. On my view ascertaining the correct answer to (a) with respect to particular scientific transitions from the past would be, rather, a matter for *historical* study. I would consider (b), however, to be the more important of the two questions. If it is answerable in the affirmative then current philosophers of science may be able to discuss, provide an account of, and give sense to, the notion of *cumulative scientific progress* with regard to scientific transitions. If so, then this would remain possible even if (a) is answered in the negative for some particular cases. That is, current philosophers of science might be able to perform the above tasks even if those who were historically involved in some scientific transitions from the past may not have been. Our approach will thus be vindicated if we can find some sort of neutral meta-language (perhaps a fragment of ordinary or semi-formalized English?) to describe common non-trivial objects dealt with by, and common non-trivial general features of, radically different, rival scientific theories.

Indeed a neutral meta-language is often analogously used in investigations of semantically interpreted formal systems or "model theory". The meta-theorems of mathematical logic serve to delimit the invariant truths about formal systems. Consider the Löwenheim-Skolem theorem. It expresses the fact that the property of having a denumerable model is invariant with respect to every consistent first-order theory. (For technical details cf. [57], pp. 65–69.) The description and proof in our meta-language of this result does not presuppose meaning invariance among the consistent first-order theories (K_1, K_2, ...); yet, it establishes and describes a shared property ("having a denumerable model") of each consistent first-order theory.

Does a similar situation prevail in physical science? Indeed it often does. In physics a comparison of statements belonging to rival theories shows that in general the theories share a number of neutral statements and expressions whose extensional meaning or reference is the same for both theories. Statements from logic and mathematics are theoretical examples of such common expressions. And in addition rival theories will frequently share statements and expressions of a more empirical or observational character. Many of the expressions of a micro-physical theory may be specific to the theory and may have extensional meanings which preclude their application to matters of familiar experience at a macroscopic level. Nevertheless, other expressions are often taken over from the language of ordinary affairs and come close to their customary everyday meanings (cf. [59], pp. 298–299). As Margenau would maintain, and correctly in my opinion, the latter refer to "bulges" on the "P-plane" ([53], p. 35). Indeed in cases of reduction Nagel feels that normally "the experiential predicates of the two sciences are the same" ([59], p. 301).

Let us now return to consider how we might describe the extensions or referents of the more empirical or observational expressions of two rival physical theories, T_1 and T_2, which are presumed radically different by radical meaning variance theorists. This must be done in a way which does not already presuppose meaning invariance between T_1 and T_2.

Surprisingly, one of Hanson's own examples fits the bill nicely ([27], p. 182). In the transition from Tycho to Kepler, Hanson feels that enough observational invariance has occurred for us to accept the following statement (due to Frank, [21], p. 231):

Our sense observation shows only that in the morning the distance between horizon and sun is increasing, but it does not tell us whether the sun is ascending or the horizon is descending... (my italics).

The italicized portion of this statement is made using a fragment of "ordinary language" which is neutral to Tycho's and Kepler's respective physical theories. *This fragment* of ordinary language is neutral to these two theories; it commits us to neither theory. Ordinary language as a whole, of course, is not so neutral; it usually speaks of the sun rising rather than of the horizon falling and is thus somewhat committed to Brahe's theory. But even here some might think it more plausible to say that it is ordinary *beliefs*, not ordinary *language*, which are committed to Brahe's theory. They might say that it is ordinary beliefs, not ordinary language, which are not here neutral. The difficulty is that there are (at least) two senses of 'ordinary language'. The first sense is close to the sense of the phrase 'ordinary classificatory scheme'. The second sense is close to the sense of the phrase 'ordinary beliefs expressed using our ordinary classificatory scheme'. In the first sense it is not all clear to me that ordinary language as a whole is committed to Brahe's rather than Kepler's theory; the laws of each theory are at least *expressible* in ordinary English, as every historian of science knows. The second sense is a different matter. One of our ordinary beliefs expressed using our ordinary classificatory scheme *is* that the sun rises; that the horizon falls is not one of these beliefs.

At any rate let us consider the above italicized *fragment* of ordinary language which Hanson does cite and accept. It purports to describe an observational situation which is invariant in this scientific transition and does this in a way which does not already *presuppose* meaning invariance between these two theories. *Accepting* the statement, however, commits Hanson, and us, to observational, and therefore meaning (extensional), invariance between the two physical theories. And in describing the invariant observational state of affairs we have used theory-neutral language: neutral, that is, between Tycho's and Kepler's respective physical theories.

Question (b) which we posed above thus seems answerable in the affirmative. That is, with regard to this particular scientific transition it seems that *we* possess enough neutral (meta) language to describe and discuss significant objects common to the ontologies of these two physical theories.

Question (a) is, of course, another question. The only languages that can be neutral between these two theories are ones that are uncommitted as to what happens when the sun rises. *Was* there such a language available to Tycho and Kepler? This becomes a historical question regarding late Medieval Latin. *Prima facie*, the answer seems to be a commitment to Tycho's side. From this, of course, it does not follow that they *could* not have constructed another language, modified Latin, or used some fragment of Latin to serve to describe non-trivial objects common to the ontologies of both of their physical theories. Neither does it follow that late Medieval Latin considered merely as a classificatory scheme was committed to Tycho's theory.

Let us consider another example of observational or experimental invariance, namely the transition from Galilean physics to Newtonian physics. Nagel thinks that the former science is reduced to the latter ([59], pp. 290–291), while Feyerabend denies it ([14], pp. 46–48). As Feyerabend ([14], p. 46) correctly notes, Galilean physics dealt with "the motion of material objects (falling stone[s], penduli, balls on an inclined planc) near the surface of the earth." Its subject matter was *terrestrial objects*. These objects were referred to by, and took on the values of, some of the terms employed by Galilean physics. It is possible to distinguish the subject matters of the Newtonian and the Galilean sciences. The latter was concerned solely with terrestrial phenomena; the former dealt also with celestial phenomena. Thus, the subject matters of the two sciences are not co-extensive. But since Newtonian physics also referred to "material objects..., near the surface of the earth", the subject matter of Galilean physics is a subset of the subject matter of Newtonian physics. The objects referred to by some of the terms employed by T_1 (Galilean physics) are also referred to by some of the terms employed by T_2 (Newtonian physics). In the transition from T_1 to T_2 there is observational invariance, and thus, (extensional) meaning invariance. As Nagel correctly notes in his discussion of the reduction of T_1 to T_2, their:

subject matters are in an obvious sense homogenous and continuous; for it is the motion of bodies and the determination of such motions that are under investigation in each case... ([59], p. 291).

Using Feyerabend's own words, we can describe the neutral observational objects in terms which are neutral to T_1 and T_2: *Both T_1 and T_2 refer to material objects such as falling stones, penduli, balls on inclined planes, etc., each of which are near the surface of the earth.* None of the terms used in this description ('falling stones', 'balls on inclined planes', etc.) are peculiar to only T_1 and T_2. Nor does the description presuppose meaning invariance between T_1 and T_2. We have not presupposed that any of the terms occurring in our description occurs with the same meaning in both T_1 and T_2; we have assumed only the ordinary English language meaning of these terms – whether or not they have the same meaning as their typographical counterparts, if any, in either T_1 or T_2. The terms in the description occur in our (and Feyerabend's) *meta-language* in which we talk *about* T_1 and T_2. Thus, in order to establish whether our description is correct and whether there exists observational invariance between T_1 and T_2, one need not presuppose meaning invariance between T_1 and T_2. And our description, indeed, is correct as every historian of science would recognize. Therefore, there is observational invariance between T_1 and T_2. And, hence, there is also (extensional) meaning invariance between Galilean and Newtonian physics. Galilean ideas of mass, acceleration, and force were applicable to all bodies located near the surface of the earth. The Newtonian idea of gravitational force included these bodies within its range and, of course, added celestial bodies as well. Margenau notes this point well:

... the [Galilean] *ideas of mass and acceleration* could be *applied to* (all my italics) a great variety of bodies, namely all those located near the surface of the earth. However, Newton's discovery of the law of universal gravitation ... was more *extensible* (his italics); it included within its range the celestial bodies. By seizing upon the *idea of a gravitational force* ... Newton provided a *concept* (all my italics) of impressive width and thereby significantly advanced the science of mechanics. ([49], p. 91)

This overlap between the ontologies of Galilean and Newtonian physics is non-trivial. The class of all terrestrial objects is not a trivial or an insignificant class. And this class comprises the overlap of the common objects dealt with by both of these physical theories. 'Terrestrial objects' is, in each theory, *a* correct answer to the question 'What moves?'. A term that illustrates non-trivial meaning invariance in both theories is, therefore, 'inertia'. In both theories it is correct to say that terrestrial objects "have inertia". There are other terms which illustrate non-trivial meaning invariance. In both theories terrestrial objects "undergo displacement". Similarly, in each theory terrestrial objects "have mass", "undergo acceleration", and "undergo velocity". There is therefore also some non-trivial meaning invariance with respect to these phrases.

Indeed our description of the neutral observational objects was also Feyerabend's. This makes it all the more surprising that Feyerabend can regard the transition from Galilean to Newtonian physics as an instance of an incommensurable change of meaning.

Our two examples of observational invariance are typical. There usually is some non-trivial observational invariance and some non-trivial (extensional) meaning invariance with respect to scientific transitions.

Meaning invariance, however, may also occur because of the invariance of abstract objects. Rival scientific theories usually share a number of highly theoretical statements and expressions whose extensional meaning or reference is the same for each theory. Mathematical statements are often examples of this. These statements are usually neutral with respect to rival scientific theories.

Ontology concerns the values available to the variables of quantification, as Quine has frequently stressed. In science, observational objects are included in the range of these values. In all current scientific theories and most past ones observational objects do not, however, exhaust this range. Abstract objects are also needed. Many things we want to say in science often force us to admit into the range of values of the variables of quantification not only objects of observation but also numbers, functions, sets, groups, points, metric spaces, and other objects of pure mathematics. As Quine correctly notes:

mathematics – not uninterpreted mathematics, but genuine set theory, logic, number theory, algebra of real and complex numbers, differential and integral calculus, and so on – is best looked upon as an integral part of science... ([66], p. 231)

Pure mathematics is usually, if not always, a part of a scientific conceptual framework. It is a commonplace that when scientific transitions occur a good deal of mathematics is usually retained. Most of the above sorts of abstract objects are, therefore, usually retained in the range of the values of the variables of quantification. From Galilean science to quantum mechanics (including their respective theories of measurement), the system of natural numbers, at least, has been retained. The abstract objects, natural numbers, have therefore been retained. In the transition from classical mechanics to general relativity Euclidean geometry was abandoned and Non-Euclidean (Riemannian) geometry was accepted. Nevertheless, the geometries used by classical mechanics and general relativity each involve some of the same abstract objects in well-defined metrical and topological senses. The transition here is from one metric geometry to another and the neutral abstract objects involved are those connected with metric (or pseudo-metric) spaces (cf. [22]). *These* neutral objects are involved in the description of the geometry of light rays in *both* theories.

The customary invariance of some abstract mathematical objects, just as the above invariance of some observational objects, ensures some (extensional) meaning invariance in scientific transitions.

The same conclusion can be approached by an indirect route. Assume that there does not exist invariance with respect to the objects dealt with by T and T'. Then when the predicates of T' are applied to the objects dealt with by T, the result will be a series of category mistakes (in Ryle's sense). They would be neither true nor false, but rather meaningless or ungrammatical or senseless, etc.; the objects of T would not be included in the range of values of the variables of quantification of T'. The resultant assertions would be neither true nor false. Just this, however, does not occur in most scientific transitions within the same science, say physics.

This we will argue in a moment. First, however, let us look at an example of the other sort of case. For example, it would make no sense to apply certain predicates and terms of quantum mechanics to entities dealt with by Keynesian economics or sociology; the resultant assertions would be neither true nor false. In particular (A), (B), and (C) would be neither true nor false, but rather meaningless or senseless:

(A)   The marginal utility of a dollar does not radiate energy continuously, but does so only in quantum jumps.

(B)   The National Character of a country does not radiate energy continuously, but does so only in quantum jumps.

(C)   The Average Taxpayer does not radiate energy continuously, but does so only in quantum jumps.

If we assert (A)–(C), it would be analogous to the category mistake John Doe makes when he "continues to think of the Average Taxpayer as a fellow-citizen" (Ryle, [70], p. 18). The situation, however, is different with:

(D)   An atomic system does not radiate energy continuously, but does so only in quantum jumps.

(D) is neither senseless, meaningless, nor a category mistake. [(D) is, in fact, true.] The reason, of course, is that an atomic system is a different sort of entity than the marginal utility of a dollar, the National Character, or the Average Taxpayer. Entities like atomic systems are dealt with by quantum mechanics; the latter entities are not. The point is that category mistakes like (A)–(C) do not occur in most scientific transitions within the same science, say physics.

Consider Hanson's example of the transition from Tycho (T) to Kepler (T'). Assume that there does not exist invariance with respect to the objects dealt with by T and T'. In particular, assume as Hanson does, that Tycho and Kepler see different suns m_0 and m_1. Thus, when the predicate of T' "is stationary relative to the earth" is applied to m_0 (the sun Tycho saw) the result should be a category mistake. This, however, is not the case. The statement that m_0 is stationary relative to the earth is meaningful and true. Historians of science, and others, do not request an explanation for the meaning of such statements nor do they claim not to understand what they are supposed to assert. When such statements are made historians of science do not exhibit the signs of lack of communication and understanding which usually accompany category mistakes.[4] And this provides at least some ground for saying that such statements are not category mistakes.

Other examples abound. The transitions from Galilean physics to Newtonian physics, and from classical physics to relativistic physics and quantum mechanics each provide numerous illustrations of cases where predicates from a new theory T' can be meaningfully applied to the objects dealt with by a previous theory T.

Consider the example of Galilean physics T and Newtonian physics T' (cf. Feyerabend, [14], pp. 46–48). One of the very reasons for the acceptance of T' in the first place was that its laws dealt with both the objects of T and celestial bodies (Margenau notes this well, cf. [49], p. 91). Newton's law of *universal* gravitation was *meant* to apply also to terrestrial objects, those dealt with by T. Indeed, Feyerabend's reasons for claiming that T cannot be reduced to T' turn on the fact that the objects dealt with by T (those close to the surface of the earth) satisfy the laws of T'; they do not undergo vertical accelerations which are constant over finite intervals of vertical distance (cf. [14], pp. 46–47). Feyerabend's discussion therefore presupposes object invariance between T and T'. He assumes that the laws of T' are true of the objects of T. And this would be impossible if there were no object invariance. If there were no object invariance category mistakes, not truths, would result from applying the laws of T' to the objects of T.

III

We have argued that there is usually some non-trivial object invariance with respect to scientific change. From this we concluded that there is usually some non-trivial (extensional) meaning invariance with respect to such change. The invariant objects were both observational and abstract. An account of how and why there exist invariant objects would prove illuminating. It should, in my opinion, proceed in good part along lines suggested by Margenau. A detailed treatment here should take account of the many relevant epistemological issues concerning the nature of perception and sensation which arise from considering the question of observation in science. This is beyond the scope of this book. We will sketch briefly only the outline of an account of observation in science. In this connection we will refer to some ideas more fully developed by Margenau. If our account is at all viable it should provide a better understanding of how and why observational invariance in fact occurs; it should also provide us with additional support for claiming that rival scientific theories share some observations, and therefore, some meaning.

Margenau presents an account of the nature of observation which lends additional support to our claim that there is usually non-trivial observational variance with respect to scientific transitions.[5] Margenau's analysis is, in my opinion, both comprehensive and precise. Following Margenau, let us denote by the term 'experience' all phases of human awareness. Let it denote every content of the mind – e.g. perceiving, thinking, feeling, willing – whether the mind be engaged in active apprehension, passive feeling, or contemplation. Sciences deal with the cognitive part of this experience, the part associated with knowledge. Two principal types of elements go into the making of scientific knowledge. These correspond to the distinction traditional philosophy has drawn between the sensory elements (acts and results of seeing, hearing, etc.) of cognitive experience and the rational elements or concepts. Representatives of the latter type are ideas such as that of a number or a function or a group. Each of these is very abstract and rationally manipulatable. The rational entities in cognitive experience, Margenau calls constructs. Taken together they make up the "C-field". The sensory elements of cognitive experience, Margenau calls the "P-domain" or "P-field".[6] They are often in the form of an immediate perception or a sensory datum or an observation (not that the latter three types of entities are necessarily identical; the point is simply that each is an element of the "P-domain").[7] The letter 'P' here may stand for 'protocol', 'perception', or 'primary'. (It should not be hastily assumed identical to any narrow logical positivist sense of 'protocol'.)

The data from the P-domain (i.e. near the P-plane) are linked to constructs via guiding relations which Margenau calls 'rules of correspondence' (cf. [49], pp. 60–69; [50], pp. 10–11). The simplest rule of correspondence is the reifying relation.[8] Other rules of correspondence are less direct. An example here is the rule by which a physicist postulates a wavelength when a particular blue light is seen. Even more indirect is the highly instrumental and refined correspondence between counter clicks (from the P-domain) of an election counter and the state function, φ, of the electron. In each case, however, and no matter how long or indirect the path, it is protocol experiences to which the constructs correspond.

Margenau notes, however, that virtually every statement about a factual experience "already involves interpretation, conceptualization, and therefore constructional elements" ([50], p. 6). He recognizes, correctly in my opinion, that a sharp boundary between sense impressions and concepts may not exist. Virtually no P-experience is pure. Virtually every protocol experience already contains an admixture of constructional elements. Margenau realizes that the simplest actual perception of a physical object is already "charged with interpretation" ([53], p. 33).[9] When language is used to record observations and experiments in an objective and significant fashion, physical objects become involved. And strictly speaking physical objects are constructs. An observation report involves more than the description of an isolated perception. It also points to other P-facts, other observations which could be made and which would be expected if certain antecedent conditions were met. Consider the observation statement, 'The opaque glass tube expanded'. This statement points up others having the form (and Hanson has noted this well cf. [27], e.g. pp. 24, 30): If α were done to X then β would result (e.g.: If the glass tube were turned around then we could observe its backside; If the glass tube were dropped then it would break). Involved in the acceptance of the first statement is the assumption that the tube has an interior and backside and that in suitable situations these too could be observed. If we accept the statement 'The opaque glass tube expanded' at t_0 then we endow the glass tube with other features which were not observed at t_0. In recognizing and allowing for these cases a position like Margenau's would preserve – where a too narrow empiricism or positivism would fail to preserve – the merits of the radical observational variance position of Hanson, Feyerabend, Kuhn, and Toulmin.

We can regard *mere* sensations or *pure* sensations (if they exist) as making up a *limiting* plane. This P-plane bounds the volume which Margenau would call the C-field. And observation and experiment, it must be admitted, do extend some distance into the C-field.[10] It is important to note, however, that this need not prevent us from saying that observations and experiments comprise a *P-domain* (as opposed to *P-plane*) with respect to a specific and particular scientific theory. Observations and experiments should be viewed as constructs (e.g. reified sense impressions) with respect to a general epistemological theory about physical objects. But with respect to a particular scientific theory they should usually be viewed as elements of the *P-domain* (as opposed to the *P-plane*). We will shortly return to this point in considering the relativization of different *P*-domains with respect to different sciences.

All this, however, does not imply that recorded observations in science are completely theory-laden as Feyerabend, Hanson, Kuhn, and Toulmin would claim. Recorded observations and experiments do bear reference to physical objects. And these strictly speaking are constructs as we have noted. But not all constructs need change when a scientific theory changes. The notion of a thing, event, physical object, or external object (each of these are constructs) *may* change when a scientific theory changes. But they usually do not. Recorded scientific experiments and observations (bulges on the *P*-plane) do refer to constructs such as these. But these constructs are not always, or even usually, specific only to the particular scientific theory undergoing test at the time; they usually do not presuppose this theory. If we appealed *only* to the constructs needed to record the observations and experiments we usually would not be able to predict what occurred. To do this we usually have to make further appeal to the special constructs and hypotheses of the particular theory we are testing. And these are usually neither used nor needed to *record* the experiments and observations.[11] Thus, the particular experiment or observation is not completely laden with, nor does it presuppose, the particular theory being tested. Kepler's use of Brahc's observation reports to test his own theory (which was incompatible with Brahe's) was, therefore, justified.

Margenau notes, and I think correctly, that strictly speaking every *science* (*not* particular theory) has its own P-domain ([50], pp. 9–10; [53], p. 34). Not every P-fact for the psychologist is a P-fact for the physicist. The principal function of P-facts is empirical verification; a P-fact which verifies Freudian psychoanalytic theory may be neutral or irrelevant with respect to quantum mechanics. Every ship needs an anchor. But the same anchor need not serve every type of ship. Battleships and rowboats, like physics and psychology, differ. This is not to say that the P-domains of different sciences need be mutually exclusive; but they usually are not coextensive. There is, however, usually significant overlap between the P-domains of different *particular* theories (e.g. Galilean mechanics and Newtonian mechanics) of any one science (e.g. physics).

A scientific theory is verified or confirmed if its consequences via the rules of correspondence meet the protocol facts and observations.[12] A necessary (but not sufficient) condition of a successful scientific theory is empirical confirmation in at least this general sense.[13] To be acceptable to science the theory must satisfy this demand, meet this standard. This (second-order) standard is both descriptive and normative. It describes *one* of the factors in fact involved in scientists' acceptance and rejection of theories; it is a standard which successful scientific theories usually *do* meet. And it is also a standard which successful theories *should* meet. At this point we see the important function and need for Margenau's analysis of observation. A P-domain is what renders empirical confirmation (or verifiability) possible for a scientific theory. The significant overlap between the P-domains of different scientific theories (e.g., Galilean and Newtonian Mechanics) within the same science (e.g. physics) is part of what renders *competition* or *rivalry* between different theories possible; each tries to account for some of the same facts. P-facts are both what confirm and what are explained by a scientific theory.[14] Without a P-domain scientific explanation could not occur (cf. [50], p. 22). Yet, isolated and unrelated P-facts alone do not confer understanding. Theory is needed. The importance of facts can be, and has been, often exaggerated.[15] Nevertheless, they are as Margenau states, "the alpha and omega of science... the starting points for theory and interpretation as well as the crucial termini conferring validity on all ideas" ([50], p. 30). Such is the indispensable function of the P-domain. Without it, scientific explanation could not take place for science would have nothing to explain. Without it, a scientific theory would not be empirically confirmable.

We have in good part followed Margenau's account of the domain of observations, the P-domain. This account deals with the nature of observation in science. It itself suggests and lends additional support to our previous claim that there exists observational invariance (and therefore meaning invariance) with respect to theoretical transitions within the same science. Margenau's characterization of the elements of the P-domain suggests that they are usually invariant with respect to scientific change.

If we have been correct in our following and elaboration of Margenau's account then scientific change is a justifiable process. It can be viewed for evaluative purposes as a deliberative process which occurs because of invariant first-level features and elements (observations, meaning) with respect to different theories.

Margenau's account further enables us to ensure the possibility of the relation "is a rival of" as used to compare different scientific theories. This we could not do on the radical observational variance account, as we argued in Chapter One. We shall say theory T is a rival of theory T' only if (a) they have the same P-domain or (b) the intersections of their respective P-domains are non-trivially non-empty. On our account (a) or (b) is usually possible and is in fact usually true for particular theories within the same general science (e.g. physics). Thus, it is usually possible for one theory to be a rival of another. To get at the relation "is better than", however, we also need to discuss other types of invariance and sharing. Unlike observational and meaning invariance, they occur at a second or meta-level. To these we now turn.

IV

I should like to suggest that comparative evaluation of rival paradigms might plausibly be viewed as a deliberative process which also occurs at a second or meta-level of discourse. Paradigm debates could then be examined and evaluated by sharable norms and a-historical standards appropriate to second-order discussion. Some sort of norms and a-historical standards are in fact held to have this function by many philosophers of science, for example, Harré, Hempel, Grünbaum, Körner, Margenau, Mehlberg, Popper, Rudner, and Suppes; and various studies of the goals and purposes of science have in fact been proposed to make such a function more precise.

Kuhn, however, suggests that sharing of second-order standards is impossible ([42], pp. 108–109). He feels that acceptance of a paradigm entails acceptance of governing standards or criteria; one can then use these to justify his acceptance of the paradigm against its rivals. Because of this Kuhn maintains that different paradigms employ different standards at the second or meta-level of scientific discourse. He concludes that each paradigm is, in effect, self-justifying. It would thus follow that paradigm debates must fail of objectivity and of the possibility of paradigm-neutral evaluation and test. Non-rational conversions would constitute the correct description of paradigm shifts within scientific communities. Scientific progress would be unattainable.

Kuhn's argument is, however, fallacious.

Kuhn does perhaps implicitly distinguish those standards or criteria which are internal to a paradigm, from those which are used to evaluate the paradigm itself. But, if any, only the former sorts of standards would seem to be "inextricably" ([42], p. 168) involved in a paradigm. To say there are internal standards inextricably involved in a paradigm is just to say that the paradigm may be understood as defining, within some scientific domain, a range of legitimate problems along with approaches to, and forms of, solution. It is to say that the paradigm provides scientists, as Kuhn says, "not only with a map but also with some of the directions essential for map-making" ([42], p. 108). However, the standards used to evaluate paradigms themselves are different in nature and function from the standards internal to a paradigm. To evaluate paradigms themselves, in terms of Kuhn's metaphor, is not to compare and evaluate different maps of a particular region with an eye to how closely they are in accord with a particular set of directions for map-making; to evaluate paradigms themselves would be, rather, as Scheffler notes ([73], p. 85), "to compare and evaluate alternative sets of directions". The latter task is a different one than the former and the standards employed are in each case different.

We can admit that a given set of mapping directions perhaps implies how the legitimacy of particular problems and proposed solutions is to be evaluated; if the problems and solutions are constructed in accordance with the directions, as they usually are, then we might say with Kuhn that "each paradigm will be shown to satisfy more or less the criteria that it dictates for itself..." ([42], pp. 108–109). In this sense one might grant that each paradigm is self-justifying. However, no set of mapping directions implies how it is itself to be evaluated in comparison with alternative sets; thus in particular, as Scheffler notes ([73], p. 85), no such set implies that it is itself superior to its alternatives. Thus, in *this* sense, it is *not* the case that each paradigm is self-justifying. Second-order standards are needed for such tasks. Paradigms themselves constitute the domain of application of such standards. These standards might then be used by philosophers of science as a philosophical rationale for the evaluation of paradigms against one another.[16] Such standards are not "dictated by", but are rather independent of, the particular paradigms to which they are applied. Assume that one's acceptance of a paradigm does entail his acceptance of some second-order standards. From this it does not follow that acceptance of different paradigms implies acceptance of different second-order standards, contrary to Kuhn ([42], p. 108), for not all relationships are one-to-one relationships. In this second-order sense then, each paradigm is *not* self-justifying, contrary to Kuhn. Kuhn's argument, therefore, has not precluded neutrality or objectivity from playing a role in paradigm evaluation. Kuhn has failed to show that paradigm differences imply evaluative differences at a second, or meta, level; he has not shown that the sharing of second-order criteria by enthusiasts of rival paradigms is impossible. He has, therefore, not demonstrated that scientific transitions consist only in non-cumulative persuasions and conversions.

My claim also is that at a second or meta-level there should be, and usually is, some invariance with respect to scientific change. Shared second or meta-level standards are needed, and used, in the business of accepting, rejecting, and evaluating rival or competing theories. They serve to regulate the choices among rival theories.

We will now briefly describe several of these guiding principles. None is absolutely invariant; each may change in time. Nevertheless, our following account of them along with some illustrations from the history of science should show that they need not change with change in scientific theory, that changes in them proceed very slowly, and that they are in fact usually invariant with respect to scientific transitions. It is through appeal to such shared standards that the "shifts of allegiance" in the scientific community usually occur.

Drawing upon the work of Margenau ([49], Chapters 5 and 6) in this area we shall briefly discuss the following: *empirical confirmation; logical fertility; extensibility; multiple connection; simplicity;* and *causality.* Our account is by no means to be taken as an exhaustive account of regulative second-order principles. Nor are the standards we present necessarily independent of one another.

Empirical confirmation:[17] A theory is confirmed if its consequences, via the rules of correspondence, meet the protocol facts and observations. Confirmation involves a circuit from the P-domain into the C-field and back again to the P-domain. We discussed this requirement in the previous section (III) in connection with the importance and need of the P-domain. Other regulative principles being equal, a theory which has sustained many circuits of empirical confirmation is usually a promising candidate for scientific acceptance. Examples of successful predictions serving as a basis for increased confidence in a theory are too numerous to mention.[18] Attempts to develop a logic of induction and recent work in confirmation theory are aimed to characterize the general notion of empirical confirmation in a precise way.[19] One of the principal problems in the philosophy of science is to determine a precise and viable characterization of this notion. The issue, in this sense, is far from settled. But, in another sense, it is quite settled. In no pernicious sense are any of the theoretical apparatuses proposed to deal with questions of inductive logic and confirmation theory theory-laden or bound by any particular scientific theory. None of the recent work in inductive logic is, for example, *inextricably* tied up with quantum mechanics. Each is invariant with respect to different scientific theories. The range of application of confirmation theory is *designed* to include *any* scientific theory. Work in inductive logic and confirmation theory is beset with problems. But in fact this has been because of technical logical difficulties and not because of problems even remotely similar to meaning variance or theory-ladenness.

Consider Goodman's work on projectibility. Will confirmation be relative to the classificatory scheme employed by a scientific theory? Yes and No. "Green" rather than "bleen" is a natural kind for us. Bearing this in mind the hypothesis that all emeralds are green is more acceptable than the hypothesis that all emeralds are bleen. In a different classificatory scheme the reverse might be true. The predicates 'is green' and 'is bleen' therefore might be said to be relative to the classificatory scheme employed by a scientific theory. But while 'is green' and 'is bleen' are so relative, 'is projectible' (in Goodman's sense) is not. The predicate 'is projectible' is not a natural kind for us. It applies to other predicates, not objects of nature. It is neither part of nor relative to the classificatory scheme employed by any current *scientific* theory. Current work in inductive logic makes it plausible to assert that if the notions of empirical confirmation or "successful prediction" or "fitting the facts" can in any wise be made precise and viable, they will be notions which in this sense are not bound by any particular scientific theory. They will be invariant and sharable second-order standards.

Logical fertility: There are different ways one might describe this demand. Margenau expresses it by saying that the constructs of a scientific theory should obey logical laws ([49], p. 82). If, for example, a scientific theory is logically inconsistent then this is grounds for its rejection. Scientific theories should be coherent. Harré also stresses this point ([30], p. 178).[20] Logical coherence is one of the regulative aims of science; other factors being equal, a fruitful scientific theory is one which promotes this aim. As we have described it this requirement is not bound to any particular scientific theory; it can be shared by, and used to evaluate, different scientific theories. It is a demand which is usually invariant with respect to changes of scientific theory; it is a commonplace that logical laws, being more abstract and general, change more slowly than scientific laws. Two valued logic is, for example, invariant with respect to the Copernican Revolution. Even with respect to quantum mechanics where many-valued logics *have* been considered, logical consistency (the law of contradiction) has not been given up.

Extensibility: Scientific theories should be extensible to as large a P-domain as possible. Other regulative principles being equal, the more extensible theory is the better theory. On this score, Newtonian mechanics which dealt with *both* terrestrial and celestial objects is a better scientific theory than Galilean mechanics which dealt only with the former objects.[21] The notion of extensibility *can*, of course, itself change and mean different things in different scientific traditions. Our point is simply that it need not and usually does not. Indeed our present notion of extensibility is applicable to virtually every past scientific theory. In discussions about science, it is, in this sense, an invariant regulative aim which is not inextricably bound to any particular scientific theory.[22]

Multiple Connection: Scientific theories should be organized and systematic. That is, the constructs used in scientific theories should be multiply connected. Scientific hypotheses should not be *ad hoc*. Hempel ([32], pp. 30–32) and Grünbaum ([25], pp. 387–388, 392–394) each stress this point well. It looms large in the latter's analysis of relativity theory. A construct or hypothesis which is *ad hoc* relative to one theory need not be *ad hoc* relative to another. But this is not to say that our ground for saying something is *ad hoc* (i.e. the concept of being *ad hoc*) is itself relative to a particular scientific theory; indeed asserting that what is *ad hoc* relative to one theory need *not* be *ad hoc* in the *same* sense relative to another theory itself presupposes an invariant ground for saying whether or not something is *ad hoc*.

Simplicity: That simplicity is often used as a regulative principle in evaluating rival scientific theories is beyond doubt. Precisely how one should characterize this notion is, however, an enormous problem. Its final characterization will, however, have to be independent of particular scientific theories if justice is to be done to the history of science. For this notion has often been appealed to in just this way as a basis for choosing among rival theories. The Copernican revolution is a case in point.[23] Ptolemic astronomy in other respects served well and predicted with quantitative precision. Copernicus, by placing the sun at the center of the planetary universe, was able to reduce the number of epicycles from 84 to about 30. This eliminated a large number of unrelated epicycles which had previously been needed to explain the same observations. Copernican astronomy was simpler than Ptolemic astronomy. In this case the operative notion of simplicity was not bound to either of the two theories; it was invariant and applicable to both. Perhaps because of its greater simplicity Copernican astronomy was also more systematic and less *ad hoc* than Ptolemic astronomy; it was more multiply connected. And this too is a factor in comparing these theories for, as Rudner aptly puts it, "System is no mere adornment of Science, it is its very heart" ([68], p. 78). Simplicity in this example is thus related to what we above called 'multiple connection'. Further, none of the current and promising investigations of simplicity[24] is bound to any particular scientific theory. They are being deliberately conducted in a way which would permit their application to more than one theory; none is, for example, *bound* to quantum theory although each will, hopefully, be applicable to this theory.

Against our position, Quine has argued that the notion of simplicity is theory-relative and not objective:

> Simplicity is not easy to define. But it may be expected, whatever it is, to be relative to the texture of a conceptual scheme. If the basic concepts of one conceptual schema are the derivative concepts of another, and vice versa, presumably one of two hypotheses could count as simpler for the one scheme and the other for the other. This being so, how can simplicity carry any peculiar presumption of objective truth? ([66], p. 242)

What Quine has in mind here is the following argument: In conceptual scheme C, the concepts used to formulate hypothesis H are basic whereas the concepts used to formulate hypothesis H' are derivative. Thus, in C, H is simpler than H'. But, in conceptual scheme C', the concepts used to formulate H are derivative whereas the concepts used to formulate H' are basic. Thus, in C', H is not simpler than H'. Therefore, simplicity is relative to the conceptual scheme employed and is not objective.

It is important to note that Quine's argument presupposes meaning invariance for H, H', and the concepts employed by C and C'. It assumes that H and H' in C can be identified with, and are the same as, H and H', respectively, in C'. It assumes that the concepts used to formulate H and H' in C can be identified with, and are the same as, those used to formulate H and H' in C'. It assumes that the *same H* is simple relative to C, but not simple relative to C'. Finally, it assumes that the *same concepts* which are basic relative to C are derivative relative to C'. Thus, Quine's argument presupposes meaning invariance at a first-level. His argument could not, therefore, be used by radical meaning variance theorists to demonstrate second-level variance, for they espouse radical meaning variance at both levels.

In any case Quine's line of thought is fallacious. In Quine's argument the criterion or ground for saying that in C, H is simpler than H' is the same as for saying that in C', H is not simpler than H'. In each case the criterion of simplicity is taken to be the employment of basic (non-derivative) concepts. In each case it is the hypothesis that employs the basic concepts which is the simpler. "Being primitive" or "being derivative" are meta-notions which are not tied to any particular first-order formal system, logical theory, or physical theory. These notions can, for example, be characterized in a way which is not bound to any first-order formal system; they are, for example, applicable to any, not to only one, first-order logical theory. Granted a particular scientific concept or scientific hypothesis which is simple relative to one scientific theory need not be simple relative to another. In noting this Quine is correct. But this is not to say that our ground for saying that something is simple (i.e. the concept of being simple) is itself relative to a particular scientific theory. Indeed Quine asserts that what is simple relative to one theory need not be simple, with respect to the *same* sense of 'simple', relative to another theory. This itself presupposes both an invariant ground for saying whether or not something is simple and an invariant sense for 'simple'. 'Simple' is assumed by Quine to have the same sense when applied to both of the conceptual schemes (C and C') of his example. The notion of simplicity in Quine's example is in this sense not relative to the conceptual scheme.

Causality: We hold with Margenau that causality is a meta-physical requirement. It demands that constructs shall be so chosen as to generate causal laws. Scientific theories should contain causal laws. The principle of causality is a good example of a regulative principle which is closely related to, if not identical with, propositions of speculative philosophy or metaphysics (e.g., the Kantian formulation as the second analogy of experience). This principle is usually a factor in regulating the scientist's construction of physical theories and his choice between them. Following Margenau ([49], p. 95) we wish to regard the principle of causality as asserting that a given state or condition of a physical system is invariably followed in time by another specifiable state or condition. With Margenau we wish to regard causality as a property of physical laws and not as a relation between single observations. This regulative principle is met by both Galilean and Newtonian mechanics and by the scientific theories of both Brahe and Kepler. The principle was not abandoned in these transitions; it continued to be employed and was in fact fulfilled by the laws of these theories. It was, therefore, invariant with respect to these scientific transitions.[25]

We have just briefly sketched some second-order standards which need not, and usually do not, change when scientific theories change. Second-order sharing can and often does occur between rival theories. These principles are primarily regulative, not constitutive.[26] Like other normative propositions they define standards of excellence. And we can choose between, and evaluate, rival scientific theories on the basis of the extent to which they meet these standards. Yet our standards are not merely normative. They also describe much of what is implicitly appealed to when scientists deem one theory more acceptable than a rival. They in fact function in science.

The relation of our second-order principles to the physical sciences might be compared to the relation of meta-mathematics to mathematics. Meta-mathematics is not intended to provide *merely* a descriptive account of how, for example, proofs are formulated in the writings of mathematicians. It is also intended to: (a) exhibit the rationale of, for example, actual mathematical proofs by revealing the logical connections underlying the successive steps; (b) provide standards for a critical appraisal of proposed proofs; (c) yield a far-reaching theory of proof, provability, decidability, and related concepts which is not bound to any particular (e.g. Euclidean geometry) mathematical theory.

Our program has analogous aims. The needed further development and characterization of our second-order principles, in addition to serving a descriptive function, would also be intended to: (a) exhibit the rationale of scientific arguments and the rationale of the choices made among rival theories by revealing the conceptual connections underlying the successive steps of these arguments and choices; (b) provide standards for a critical appraisal of proposed scientific arguments; (c) yield a comprehensive theory which blends inductive logic, confirmation, simplicity, causality, and related concepts and which is not bound to any particular scientific theory.

We have sketched only the barest outline of such a program which is sufficient for our purposes here. Obviously, much further work is needed for the articulation of such second-order standards. Yet, even from the little we have said it is clear that our second-order principles and standards need not, and usually do not, change when scientific theories change. None of these principles and standards is paradigm relative; each can be shared. They are neither bound to, nor do they presuppose, any particular scientific theory. They can be, and are, used to evaluate different scientific theories against one another. Fruitful scientific constructs, hypotheses, and theories are those which contribute to the fulfillment of at least these second-order aims.

One of the merits of the approach of radical meaning variance theorists is that it calls our attention to the fact that non-factual criteria *are* involved in the acceptance or rejection of theories. "Fitting the facts" is indeed not the only criterion involved.[27] Paradigm disputes may have sometimes in fact occurred because different parties have employed different regulative standards. Noting this may remove some of what is still unexplained in trying to fully account for why some particular paradigm disputes ever took place. Our point, however, is that holding different scientific theories does not in itself commit one to holding different regulative principles. Our point is that regulative principles need not, and usually do not, change when scientific theories change.

On our view objectivity is an ideal which is in fact employed in scientific practice. Science is a systematic public enterprise that can be justified by reference to second-level standards, logic, and empirical fact. None of the latter is completely theory-laden; none of the latter need presuppose any particular scientific theory. Part of the purpose of science is to formulate truths about the natural world in a simple, comprehensive, systematic, and intelligible way in which nature becomes explainable, predictable, and controllable. The success of any particular scientific theory is a measure of how far, and how successfully it contributes to the realization of the general second-level aims of science. If significant first-level sharing occurs between T and a competitor T' and if T takes us further in the direction of the above second-level aims than T', then T should be accepted and T' rejected.

NOTES

[1] As we shall see, Margenau would not regard this as the whole story. His position is more comprehensive than positivism. Quite rightly, he would hold that there are second-order (or meta-level) principles involved here in addition to "fitting the facts".
[2] This statement is accurate enough for my present purposes, since my argument in this section requires only that extension be a first step toward an account of meaning. That is, it requires only that a significant part of the meaning of a term is its extension.
[3] Such an objection prompts Alexander to approach the question of meaning invariance in a different way than we will. (Cf. [3], p. 83): "We are not directly interested in the question 'Do two people ever see the same thing?' since we cannot discover whether they are, or are not, seeing the same thing except by discovering whether they do, or can, accept the same description. It must be stressed that whether two forms of words constitute 'the same description' depends not merely upon the forms of words but also upon the particular meanings which are being attached to them on the relevant occasions."
[4] For some good illustrations of category mistakes and the types of behavior which usually accompany them cf.: Ryle, [70], pp. 16–18.

[5] Cf.: [49], pp. 13, 25, 44, 48–50, 55–56, 58–61, 64–69, 72–73, 103–107; [50], pp. 2, 4–10, 14–19, 22–23, 28–30, 66; [53].

[6] Representatives here are "what are commonly known as *facts*" ([50], p. 4).

[7] The *P*-domain serves a controlling function in our choice of theories. It is *P*-experience "against which scientific theories and all other kinds of conjectures are ultimately tested" ([50], p. 5). We shall shortly return to, and elaborate upon, this point.

[8] An example here is the rule which associates with the "various ephemeral aspects of the visual tree" the public or external object, "tree" ([49], p. 58).

[9] Margenau notes that "when language is used to communicate the occurrence of a perception, a passage to the world of symbolism and therefore construction has already taken place. Therefore, the word protocol will apply not to mere sensations devoid of meanings, which if possible at all are scientifically uninteresting, but to slightly more elaborate complexes such as *reified* sense impressions and to refined sense impressions" ([53], pp. 33–34).

He notes that in physics "A direct observation of a physical phenomenon is a sense datum, but it is also a little more since, when recorded in objectively significant fashion, it already bears reference to physical objects which strictly are constructs" ([53], p. 34). These points comprise, I think, a good part of what (e.g.) Hanson was driving at. These points *are* important and radical observational variance theorists have been correct to concern themselves with them.

[10] Margenau expresses the matter as follows: "a *P*-experience arises on and refers to that plane [the *P*-plane] but extends some distance into the *C*-field. Figuratively, it may be convenient to think of it as a "bulge" on the *P*-plane" ([53], p. 35). But it should be noted here that among the sciences Margenau thinks that "Physics *comes closest* to identifying it [its protocol facts or *P*-domain] with traditional sense data, i.e., with John Locke's simple ideas, Immanuel Kant's Wahrnehmungen, or the contents of Rudolf Carnap's early protocol sentences" ([53], p. 34, my italics).

[11] The experiments or factual matters which led to the breakdown of classical physics could be and were *described* without appeal to the concepts (constructs) of quantum mechanics, such as φ, the state function of the electron, although these concepts were later needed for the *explanation* of the experimental results (cf. Margenau, [49], pp. 67–69). For a good discussion of these experiments – viz., those which brought about the breakdown of classical mechanics – cf. Margenau, [49], Chapter 16.

[12] Margenau has described the general features of the matter in a number of publications; cf. especially: [53], pp. 31–32; [49], Chapter 6.

[13] We will discuss the other demands and standards which a successful scientific theory does and should meet in the next Section (IV).

[14] Cf. Margenau: "If science gives up the premise of a P-plane, it loses the very basis of its competence" ([50], p. 10).

[15] For a critique of some narrow empiricisms which do just this, cf. Margenau, [50], pp. 27–30 and [49], pp. 18–20.

[16] For a good discussion of second-order standards with examples of their actual application cf. Margenau, [49], Chapter 5, "Metaphysical Requirements On Constructs", pp. 75–101. Also relevant here is Rudner's survey of the literature on the notion of simplicity and his discussion of its role as a second-order requirement and standard which successful theories should meet [68]. We shall soon consider the relevance of their views.

[17] The regulative function of the criterion of empirical confirmation has also been noted by many other philosophers, cf.: Harré, [30], Chapter 6, e.g., p. 144; Hempel,

[33], p. 116; Purtill, [62], p. 54; Rudner, [68], pp. 75–76; and Scheffler, [73], pp. 8–10.

[18] Cf. Margenau, [49], p. 104, for some good examples here.

[19] For a good overview of some recent work in this area, cf. Hempel, [33].

[20] Harré's general discussion of second-level criteria ([30], Chapter 6, 'Fitting the Facts' and Chapter 7, 'Non-Factual Criteria') makes some distinctions analogous to those made by Margenau ([49], Chapter 6, 'Empirical Confirmation' and Chapter 5, 'Metaphysical Requirements On Constructs'). With respect to second-level standards Margenau's distinction between 'Empirical Confirmation' and 'Metaphysical Requirements On Constructs' roughly corresponds to Harré's distinction between the criterion of 'Fitting the Facts' and the 'Non-Factual Criteria'.

[21] And if Margenau is correct, as I think he may be, quantum mechanics is, on this score, better than classical mechanics ([49], pp. 383–388; [52], pp. 349–350, 356; the latter reference is explicitly concerned with Feyerabend's counter-claim).

[22] For further discussion with several non-trivial illustrations from the history of science cf. Margenau [49], pp. 90–94.

[23] With regard to this example, cf. Margenau, [49], pp. 96–97; [53], pp. 32–33.

[24] For a good introduction to some recent work in this area, cf. Rudner, [68].

[25] If Margenau is correct in his description of the principle of causality and in his analysis of its application to quantum theory ([49], Chapter 19, esp. pp. 418–422; [48]), as I think he is, then this is a regulative principle which is also met by both classical and quantum mechanics. If he is correct here then the principle has not been abandoned in this scientific transition; it would still be employed and in fact fulfilled; it would, therefore, be invariant with respect to this transition also. Since Margenau takes causality in Kantian fashion to be a relation between states rather than, as with Hume, a relation between single observations, quantum theory presumably poses no difficulty. We have not presented detailed arguments for his position on this point; rather we have indicated only where they can be found. A detailed account here, along with the issues it would raise with regard to quantum mechanics, is beyond the scope of this book.

[26] For a good discussion which emphasizes the *regulative* function of second-level principles, cf. Körner, [40].

[27] Cf. in this light Kuhn: "No process yet disclosed by the historical study of scientific development at all resembles the methodological stereotype of falsification by direct comparison with nature" ([42], p. 77).

BIBLIOGRAPHY

[1] Achinstein, P., 'On the Meaning of Scientific Terms', *The Journal of Philosophy* **61** (1964) 497–509.

[2] Achinstein, P., *Concepts of Science*, The Johns Hopkins Press, Baltimore, 1968.

[3] Alexander, P., *Sensationalism and Scientific Explanation*, Routledge and Kegan Paul, London, 1963.

[4] Alston, W. P., *Philosophy of Language,* Prentice-Hall, Englewood Cliffs, 1964.

[5] Austin, J. L., 'Other Minds', in *Philosophical Papers*, Oxford University Press, Oxford, 1961.

[6] Austin, J. L., *How to do Things with Words*, Oxford University Press, London, 1962.

[7] Bohr, N., 'On the Constitution of Atoms and Molecules', *Philosophical Magazine* **26** (1913).

[8] Carnap, R., 'Testability and Meaning', *Philosophy of Science* **3** (1936) 419–471, and **4** (1937) 1–40.

[9] Cole, M., Gay, J., Glick, J., and Sharp, D., 'Linguistic Structure and Transposition', *Science* **164** (1969) 90–91.

[10] Daniels, C. B., 'Colors and Sensations: Or How to Define a Pain Ostensively', *American Philosophical Quarterly* **4** (1967) 231–237.

[11] Daniels, C. B., 'Sense Data', privately distributed (1968) and submitted for publication.

[12] Duhem, P., *La théorie physique*, Paris 1914.

[13] Einstein, A., 'Autobiographical Notes', in *Albert Einstein: Philosopher-Scientist* (ed. by P. A. Schilpp), Tudor Publishing Co., New York, 1949.

[14] Feyerabend, P. K., 'Explanation, Reduction, and Empiricism', in *Minnesota Studies in the Philosophy of Science* (ed. by H. Feigl and G. Maxwell), *Scientific Explanation, Space, and Time*, vol. III, pp. 28–97, University of Minnesota Press, Minneapolis, 1962.

[15] Feyerabend, P. K., 'How to Be a Good Empiricist: A Plea for Tolerance in Matters Epistemological', in *The Delaware Seminar in Philosophy of Science* (ed. by B. Baumrin), vol. 2, pp. 3–39, Interscience, New York, 1963.

[16] Feyerabend, P. K., 'Problems of Empiricism', in *Beyond the Edge of Certainty*, (ed. by R. Colodny), pp. 145–260, Prentice-Hall, Englewood Cliffs, 1965.

[17] Feyerabend, P. K., 'On the "Meaning" of Scientific Terms', *The Journal of Philosophy* **62** (1965) 266–274.

[18] Feyerabend, P. K., 'Reply to Criticism', in *Boston Studies in the Philosophy of Science* (ed. by R. S. Cohen and M. W. Wartofsky), vol. 2, pp. 223–261, Humanities Press, New York, 1965.

[19] Feyerabend, P. K., 'Consolations for the Specialist', in *Criticism and the Growth of Knowledge* (ed. by I. Lakatos and A. Musgrave), pp. 197–230, Cambridge University Press, Cambridge, 1970.

[20] Fine, A. I., 'Consistency, Derivability, and Scientific Change', *The Journal of Philosophy* **64** (1967) 231–240.

[21] Frank, P., *Modern Science and its Philosophy*, Harvard University Press, Cambridge, 1949.

[22] Fréchet, M., 'Sur quelques points du calcul fonctionnel', *Rendiconti del Circolo Matematico* **22** (1906) 1–74.

[23] Goodman, N., *The Structure of Appearance*, Harvard University Press, Cambridge, 1938.

[24] Goodman, N., 'On Likeness of Meaning', in *Semantics and the Philosophy of Language* (ed. by L. Linsky), pp. 67–74, University of Illinois Press, Urbana, 1952.

[25] Grünbaum, A., *Philosophical Problems of Space and Time*, Alfred A. Knopf, New York, 1963.

[26] Halliday, D. and Resnick, R., *Physics*, Wiley and Sons, Inc., New York, 1962.

[27] Hanson, N. R., *Patterns of Discovery*, The University Press, Cambridge, 1958.

[28] Hanson, N. R., *The Concept of the Positron*, The University Press, Cambridge, 1963.

[29] Harman, G., 'Quine on Meaning and Existence, I', *Review of Metaphysics* **21** (1967) 124–151.

[30] Harré, R., *An Introduction to the Logic of the Sciences*, Macmillan and Co., London, 1963.

[31] Hempel, C. G., *Fundamentals of Concept Formation in Empirical Science*, University of Chicago Press, Chicago, 1952.

[32] Hempel, C. G., *Philosophy of Natural Science*, Prentice-Hall, Inc., Englewood Cliffs, 1966.

[33] Hempel, C. G., 'Recent Problems of Induction', in *Mind and Cosmos* (ed. by R. G. Colodny), 112–134, University of Pittsburgh Press, Pittsburgh, 1966.

[34] Hesse, M., 'A New Look at Scientific Explanation', *Review of Metaphysics* **17** (1963) 98–108.

[35] Hesse, M., 'Fine's Criteria of Meaning Change', *The Journal of Philosophy* **65** (1968) 46–52.

[36] Holton, G. and Roller, D., *Foundations of Modern Physical Science*, Addison-Wesley, Reading, Mass., 1958.

[37] Kepleri, J., *Astronomi Opera Omnia* (ed. by C. Frisch), 8 vols., Frankfurt and Erlangen 1858.

[38] Kershner, R. B. and Wilcox, L. R., *The Anatomy of Mathematics*, The Ronald Press, New York, 1950.

[39] Kordig, C. R., 'On Prescribing Description', *Synthese* **18** (1968) 459–461.

[40] Körner, S., 'On Philosophical Arguments in Physics', in *The Structure of Scientific Thought* (ed. by E. H. Madden), pp. 106–110, Houghton Mifflin Company, Boston, 1960.

[41] Kuhn, T. S., *The Copernican Revolution*, Vintage Books, New York, 1957.

[42] Kuhn, T. S., *The Structure of Scientific Revolutions*, University of Chicago Press, Chicago, 1962.

[43] Kuhn, T. S., 'Logic of Discovery or Psychology of Research?', to be published in *The Philosophy of K. R. Popper* (ed. by P. A. Schilpp).

[44] Lavoisier, A., 'Experiments of Animals and on the Changes which happen to Air in its Passage through their Lungs', in *Readings in the Literature of Science*, (ed. by W. C. Dampier and M. Dampier), pp. 88–92, Harper Torchbooks, New York, 1959.

[45] Lavoisier, A., 'Memoir on the Combustion of Candles in Atmospheric Air and in Respirable Air', in *Readings in the Literature of Science* (ed. by W. C. Dampier and M. Dampier), p. 92, Harper Torchbooks, New York, 1959.

[46] Lewis, C. I., *Mind and the World Order*, Dover, New York, 1929.

[47] Lindsay, R. B. and Margenau, H., *Foundations of Physics*, Dover, New York, 1957.

[48] Margenau, H., 'Meaning and Scientific Status of Causality', in *Philosophy of Science* (ed. by A. Danto and S. Morgenbesser), pp. 435–449, The World Publishing Company, New York, 1960.

[49] Margenau, H., *The Nature of Physical Reality*, McGraw-Hill, New York, 1950.

[50] Margenau, H., *Open Vistas*, Yale University Press, New Haven, 1961.

[51] Margenau, H., 'Is the Mathematical Explanation of Physical Data Unique?', in *Logic, Methodology and Philosophy of Science* (ed. by E. Nagel, P. Suppes, and A. Tarski), pp. 348–355, Stanford University Press, Stanford, 1962.

[52] Margenau, H., 'The Philosophical Legacy of the Quantum Theory', in *Mind and Cosmos* (ed. by R. Colodny), pp. 330–356, University of Pittsburgh Press, 1966.

[53] Margenau, H., 'What Is a Theory', in *The Structure of Economic Science* (ed. by S. R. Krupp), pp. 25–38, Prentice-Hall, Englewood Cliffs, 1966.

[54] Mason, S. F., *A History of the Sciences*, Collier Books, New York, 1962.

[55] Mates, B., 'Synonymity', in *Semantics and the Philosophy of Language* (ed. by L. Linsky), pp. 111–136, University of Illinois Press, Urbana, 1952.

[56] Mehlberg, H., 'The Theoretical and Empirical Aspects of Science', in *Logic, Methodology and Philosophy of Science* (ed. by E. Nagel, P. Suppes, and A. Tarski), pp. 275–284, Stanford, 1962.

[57] Mendelson, E., *Introduction to Mathematical Logic*, Van Nostrand, Princeton, 1964.

[58] Morris, C., *Foundations of the Theory of Signs*, University of Chicago Press, Chicago, 1938.

[59] Nagel, E., 'The Meaning of Reduction in the Natural Sciences', in *Philosophy of Science* (ed. by A. Danto and S. Morgenbesser), pp. 288–312, World Publishing Co., Cleveland, 1960.

[60] Polanyi, M., *Personal Knowledge*, The University of Chicago Press, Chicago, 1962.

[61] Popper, K. R., *The Logic of Scientific Discovery*, Harper and Row, New York, 1965.

[62] Purtill, R. L., 'Kuhn on Scientific Revolutions', *Philosophy of Science* **34** (1967) 53–58.

[63] Putnam, H., 'How Not to Talk about Theories', in *Boston Studies in the Philosophy of Science* (ed. by R. S. Cohen and M. W. Wartofsky), vol. 2, pp. 205–222, Humanities Press, New York, 1965.

[64] Quine, W. V., *From a Logical Point of View*, Harvard University Press, Cambridge, 1953.

[65] Quine, W. V., *Word and Object*, M.I.T. Press, Cambridge, 1960.

[66] Quine, W. V., *The Ways of Paradox and Other Essays*, Random House, New York, 1966.

[67] Reichenbach, H., *Experience and Prediction*, University of Chicago Press, Chicago, 1938.

[68] Rudner, R., 'An Introduction to Simplicity', in *Philosophical Problems of Natural Science* (ed. by D. Shapere), pp. 75–81, Macmillan Company, New York, 1965.

[69] Ryle, G., 'Systematically Misleading Expressions', in *Logic and Language* (*First and Second Series*), (ed. by A. Flew), Doubleday, New York, 1965.
[70] Ryle, G., *The Concept of Mind*, Barnes and Noble, New York, 1949.
[71] Ryle, G., *Dilemmas*, University Press, Cambridge, 1954.
[72] Ryle, G., 'The Theory of Meaning', in *Philosophy and Ordinary Language* (ed. by C. E. Caton), University of Illinois Press, Urbana, 1963.
[73] Scheffler, I., *Science and Subjectivity*, The Bobbs-Merrill Company, Indianapolis, 1967.
[74] Sellars, W., 'Empiricism and the Philosophy of Mind', in *Science, Perception and Reality*, pp. 127–196, Routledge and Kegan Paul, London, 1963.
[75] Shapere, D., 'Space, Time, and Language – An Examination of Some Problems and Methods of the Philosophy of Science', in *The Delaware Seminar in Philosophy of Science* (ed. by B. Baumrin), vol. 2, pp. 139–170, Interscience, New York, 1963.
[76] Shapere, D., 'The Structure of Scientific Revolutions', *Philosophical Review* **73** (1964) 383–94.
[77] Shapere, D., *Philosophical Problems of Natural Science*, Macmillan, New York, 1965.
[78] Shapere, D., 'Meaning and Scientific Change', in *Mind and Cosmos* (ed. by R. Colodny), pp. 41–85, University of Pittsburgh Press, Pittsburgh, 1966.
[79] Shapere, D., 'Notes Toward a Post-Positivistic Interpretation of Science', in *The Legacy of Logical Positivism* (ed. by P. Achinstein and S. Barker), Johns Hopkins University Press, Baltimore, 1969.
[80] Sklar, L., 'Types of Inter-Theoretic Reduction', *British Journal for the Philosophy of Science* **18** (1967) 109–124.
[81] Smart, J. J. C., 'Conflicting Views about Explanation', in *Boston Studies in the Philosophy of Science*, (ed. by R. S. Cohen and M. W. Wartofsky), vol. 2, pp. 157–169, Humanities Press, New York, 1965.
[82] Toulmin, S., *The Philosophy of Science*, Harper and Row, New York, 1953.
[83] Toulmin, S., *Foresight and Understanding*, Harper and Row, New York, 1961.
[84] Toulmin, S., 'Conceptual Revolutions in Science', *Synthese* **17** (1967) 75–91.
[85] Toulmin, S., 'Reply', *Synthese* **18** (1968) 462–463.
[86] Van Fraassen, B. C., 'Presupposition, Implication, and Self-reference', *Journal of Philosophy* **65** (1968) 136–152.
[87] Whorf, B. L., 'Science and Linguistics', in *Language, Thought and Reality* (ed. by J. B. Carroll), M.I.T. Press, Cambridge, 1957.
[88] Wittgenstein, L., *Philosophical Investigations*, Macmillan, New York, 1958.

SYNTHESE LIBRARY

Monographs on Epistemology, Logic, Methodology,
Philosophy of Science, Sociology of Science and of Knowledge, and on the
Mathematical Methods of Social and Behavioral Sciences

Editors:
Donald Davidson (Rockefeller University and Princeton University)
Jaakko Hintikka (Academy of Finland and Stanford University)
Gabriel Nuchelmans (University of Leyden)
Wesley C. Salmon (Indiana University)

‡Milič Čapek, *Bergson and Modern Physics. A Reinterpretation and Re-evaluation.*
Boston Studies in the Philosophy of Science. Volume VII (ed. by Robert S. Cohen
and Marx W. Wartofsky). 1971, XI + 414 pp. Dfl. 70,—

‡Joseph D. Sneed, *The Logical Structure of Mathematical Physics.* 1971, XV + 311 pp.
Dfl. 70,—

‡Jean-Louis Krivine, *Introduction to Axiomatic Set Theory.* 1971, VII + 98 pp.
Dfl. 28,—

‡Risto Hilpinen (ed.), *Deontic Logic: Introductory and Systematic Readings.* 1971,
VII + 182 pp. Dfl. 45,—

‡Evert W. Beth, *Aspect of Modern Logic.* 1970, XI + 176 pp. Dfl. 42,—

‡Paul Weingartner and Gerhard Zecha, (eds.), *Induction, Physics, and Ethics.*
Proceedings and Discussions of the 1968 Salzburg Colloquium in the Philosophy of
Science. 1970, X + 382 pp. Dfl. 65,—

‡Rolf A. Eberle, *Nominalistic Systems.* 1970, IX + 217 pp. Dfl. 42,—

‡Jaakko Hintikka and Patrick Suppes, *Information and Inference.* 1970, X + 336 pp.
Dfl. 60,—

‡Karel Lambert, *Philosophical Problems in Logic. Some Recent Developments.* 1970,
VII + 176 pp. Dfl. 38,—

‡P. V. Tavanec (ed.), *Problems of the Logic of Scientific Knowledge.* 1969, XII + 429 pp.
Dfl. 95,—

‡Robert S. Cohen and Raymond J. Seeger (eds.), *Boston Studies in the Philosophy of*
Science. Volume VI: *Ernst Mach: Physicist and Philosopher.* 1970, VIII + 295 pp.
Dfl. 38,—

‡Marshall Swain (ed.), *Induction, Acceptance, and Rational Belief.* 1970, VII + 232 pp.
Dfl. 40,—

‡Nicholas Rescher *et al.* (eds.), *Essays in Honor of Carl G. Hempel. A Tribute on the*
Occasion of his Sixty-Fifth Birthday. 1969, VII + 272 pp. Dfl. 46,—

‡Patrick Suppes, *Studies in the Methodology and Foundations of Science. Selected*
Papers from 1951 to 1969. 1969, XII + 473 pp. Dfl. 70,—

‡JAAKKO HINTIKKA, *Models for Modalities. Selected Essays.* 1969, IX + 220 pp.
Dfl. 34,—

‡D. DAVIDSON and J. HINTIKKA (eds.), *Words and Objections: Essays on the Work of W. V. Quine.* 1969, VIII + 366 pp.
Dfl. 48,—

‡J. W. DAVIS, D. J. HOCKNEY and W. K. WILSON (eds.), *Philosophical Logic.* 1969, VIII + 277 pp.
Dfl. 45,—

‡ROBERT S. COHEN and MARX W. WARTOFSKY (eds.), *Boston Studies in the Philosophy of Science.* Volume V: *Proceedings of the Boston Colloquium for the Philosophy of Science 1966/1968.* 1969, VIII + 482 pp.
Dfl. 58,—

‡ROBERT S. COHEN and MARX W. WARTOFSKY (eds.), *Boston Studies in the Philosophy of Science.* Volume IV: *Proceedings of the Boston Colloquium for the Philosophy of Science 1966/1968.* 1969, VIII + 537 pp.
Dfl. 69,—

‡NICHOLAS RESCHER, *Topics in Philosophical Logic.* 1968, XIV + 347 pp.   Dfl. 62,—

‡GÜNTHER PATZIG, *Aristotle's Theory of the Syllogism. A Logical-Philological Study of Book A of the Prior Analytics.* 1968, XVII + 215 pp.
Dfl. 45,—

‡C. D. BROAD, *Induction, Probability, and Causation. Selected Papers.* 1968, XI + 296 pp.
Dfl. 48,—

‡ROBERT S. COHEN and MARX W. WARTOFSKY (eds.), *Boston Studies in the Philosophy of Science.* Volume III: *Proceedings of the Boston Colloquium for the Philosophy of Science 1964/1966.* 1967, XLIX + 489 pp.
Dfl. 65,—

‡GUIDO KÜNG, *Ontology and the Logistic Analysis of Language. An Enquiry into the Contemporary Views on Universals.* 1967, XI + 210 pp.
Dfl. 38,—

*EVERT W. BETH and JEAN PIAGET, *Mathematical Epistemology and Psychology.* 1966, XXII + 326 pp.
Dfl. 58,—

*EVERT W. BETH, *Mathematical Thought. An Introduction to the Philosophy of Mathematics.* 1965, XII + 208 pp.
Dfl. 32,—

‡PAUL LORENZEN, *Formal Logic.* 1965, VIII + 123 pp.
Dfl. 22,—

‡GEORGES GURVITCH, *The Spectrum of Social Time.* 1964, XXVI + 152 pp. Dfl. 20,—

‡A. A. ZINOV'EV, *Philosophical Problems of Many-Valued Logic.* 1963, XIV + 155 pp.
Dfl. 28,—

‡MARX W. WARTOFSKY (ed.), *Boston Studies in the Philosophy of Science.* Volume I: *Proceedings of the Boston Colloquium for the Philosophy of Science, 1961–1962.* 1963, VIII + 212 pp.
Dfl. 22,50

‡B. H. KAZEMIER and D. VUYSJE (eds.), *Logic and Language. Studies dedicated to Professor Rudolf Carnap on the Occasion of his Seventieth Birthday.* 1962, VI + 246 pp.
Dfl. 32,50

*EVERT W. BETH, *Formal Methods. An Introduction to Symbolic Logic and to the Study of Effective Operations in Arithmetic and Logic.* 1962, XIV + 170 pp.   Dfl. 30,—

*HANS FREUDENTHAL (ed.), *The Concept and the Role of the Model in Mathematics and Natural and Social Sciences. Proceedings of a Colloquium held at Utrecht, The Netherlands, January 1960.* 1961, VI + 194 pp.
Dfl. 30,—

‡P. L. R. GUIRAUD, *Problèmes et méthodes de la statistique linguistique.* 1960, VI + 146 pp.
Dfl. 22,50

*J. M. BOCHEŃSKI, *A Precis of Mathematical Logic.* 1959, X + 100 pp.   Dfl. 20,—

SYNTHESE HISTORICAL LIBRARY

Texts and Studies
in the History of Logic and Philosophy

Editors:

N. KRETZMANN (Cornell University)
G. NUCHELMANS (University of Leyden)
L. M. DE RIJK (University of Leyden)

‡KARL WOLF and PAUL WEINGARTNER (eds.), *Ernst Mally: Logische Schriften.* 1971,
X + 340 pp.   Dfl. 80,—

‡LEROY E. LOEMKER (ed.), *Gottfried Wilhelm Leibniz: Philosophical Papers and Letters.*
A Selection Translated and Edited, with an Introduction. 1969, XII + 736 pp.
Dfl. 125,—

‡M. T. BEONIO-BROCCHIERI FUMAGALLI, *The Logic of Abelard.* Translated from the
Italian. 1969, IX + 101 pp.   Dfl. 25,—

Sole Distributors in the U.S.A. and Canada:

*GORDON & BREACH, INC., 150 Fifth Avenue, New York, N.Y. 10011
‡HUMANITIES PRESS, INC., 303 Park Avenue South, New York, N.Y. 10010

PHILOSOPHY DEPARTMENT LIBRARY
Waggener Hall 312
University of Texas
Austin, Texas 78712